STAMS® SERIES
B

STRATEGIES TO ACHIEVE MATHEMATICS SUCCESS

- ☐ PROVIDES INSTRUCTIONAL ACTIVITIES FOR 12 MATHEMATICS STRATEGIES
- ☐ USES A STEP-BY-STEP APPROACH TO ACHIEVE MATHEMATICS SUCCESS
- ☐ PREPARES STUDENTS FOR SELF-ASSESSMENT IN MATHEMATICS COMPREHENSION

CURRICULUM ASSOCIATES®, INC.

Acknowledgments

Product Development by Chameleon Publishing Services

Revision (2006) by Robert G. Forest, EdD.

ISBN 978-0-7609-3648-1

North Billerica, MA 01862

15 14 13 12 11 10 9 8 7 6 5 4

TABLE OF CONTENTS

PART ONE: Think About Number Sense

WHAT DO YOU KNOW ABOUT NUMBERS?

Each digit in a number has place value, such as ones, tens, and hundreds.

The value of a digit depends on its place in a number.

A number may be written in digits or in words.

Use number words like first, second, and sixth to show the position of something in a row or line.

▶ Look at the chart. Then answer the questions or follow the directions.

hundreds (100)	tens (10)	ones (1)
2	4	6

a. What number does the chart show? ____________

b. What digit is in the ones column? ____________

c. What digit is in the tens column? ____________

d. What digit is in the hundreds column? ____________

e. Write the number in words. ________________________________

f. The number can also be shown as ____00 + ____0 +____.

▶ Write 3 things about the number 327.

____

____

____

You just reviewed information about place value and writing numbers.

WHAT MORE DO YOU KNOW ABOUT NUMBERS?

When you count, you tell what number comes before or after a number.

▶ Look at each group of numbers. Write the number that comes before or after a number.

a. 16, 17, ___, 19

b. ___, 30, 31, 32

c. 97, 98, 99, ___

d. ___, 110, 111, 112

> You just reviewed the order of counting numbers.

▶ Look at each row of items. Answer each question about the position of the colored item in the row.

a. **Start**     

In what place is the colored fish? ______________________

b. **Start**      

In what place is the colored beach ball? ______________________

> You just reviewed how to use a number word to tell the place of an item in a row.

> Alone, write two problems using the number word *fourth* in the first problem and the number word *sixth* in the second problem. Write a question about each problem. When finished, give your problems to your partner to solve. You solve your partner's problems. Together, talk over the solutions.

PART TWO: Learn About Number Sense

Study the place-value chart that Theo's teacher made. It shows how many people visited the second-grade science fair. As you study, think about the place value of each digit.

hundreds (100)	tens (10)	ones (1)
6	7	5

The chart shows the number 675.
The number 675 is written as six hundred seventy-five.
The number 675 has three digits.
The number 675 has 6 hundreds, 7 tens, and 5 ones.
The number 675 can be shown as 600 + 70 + 5.

Look at some other numbers and how they can be written.
The number 59 has 5 tens and 9 ones.
The number 280 can be shown as 200 + 80 + 0.
The number 364 is written as three hundred sixty-four.

You use **number sense** when you think about the place value of each digit in a number.

- Each digit in a number has a place value, such as ones, tens, or hundreds.
- The value of a digit depends on its place in a number.
- A number can be written in digits or in words.

Theo made his own place-value chart. Theo's chart shows the number of ants in his ant farm. Study Theo's chart. Think about the number the chart shows. Think about the place value of each digit in the number. Then do Numbers 1 through 4.

hundreds (100)	tens (10)	ones (1)
8	4	5

1. What number does the chart show?

Ⓐ 854
Ⓑ 845
Ⓒ 548
Ⓓ 458

2. What is the place value of the 4 in the number 845?

Ⓐ 5 ones
Ⓑ 4 ones
Ⓒ 4 tens
Ⓓ 4 hundreds

3. Which of these shows how the number 845 can be written?

Ⓐ eight hundred forty-five
Ⓑ eighty-five
Ⓒ eight hundred nine
Ⓓ eighty

4. Which of these shows the number 845?

Ⓐ 400 + 50 + 8
Ⓑ 800 – 40 – 5
Ⓒ 800 + 40 + 5
Ⓓ 500 + 80 + 4

Talk about your answers to questions 1–4.
Tell why you chose the answers you did.

PART THREE: Check Your Understanding

Remember: You use number sense when you think about the place value of each digit in a number.

- Each digit in a number has a place value, such as ones, tens, or hundreds.
- The value of a digit depends on its place in a number.
- A number can be written in digits or in words.

Solve this problem. As you work, ask yourself, "What are the ways that numbers can be written?"

5. Theo attends Willow Middle School. The lunchroom at the school serves two hundred ninety-six hot lunches to students each day. What number shows how many hot lunches are served each day?

Ⓐ 26

Ⓑ 29

Ⓒ 209

Ⓓ 296

Solve another problem. As you work, ask yourself, "What does the place of each digit in a number tell me about its place value?"

6. There are 579 students in the Willow Middle School. What is the place value of the 7 in the number 579?

Ⓐ 9 ones

Ⓑ 7 ones

Ⓒ 7 tens

Ⓓ 7 hundreds

Look at the answer choices for each question.
Read why each answer choice is correct or not correct.

5. Theo attends Willow Middle School. The lunchroom at the school serves two hundred ninety-six hot lunches to students each day. What number shows how many hot lunches are served each day?

Ⓐ 26

This answer is not correct because 26 is twenty-six.

Ⓑ 29

This answer is not correct because 29 is twenty-nine.

Ⓒ 209

This answer is not correct because 209 is two hundred nine.

● 296

This answer is correct because 296 is written two hundred ninety-six.

6. There are 579 students in the Willow Middle School. What is the place value of the 7 in the number 579?

Ⓐ 9 ones

This answer is not correct because 9 ones equals 9, and the 9 in 579 is in the ones place.

Ⓑ 7 ones

This answer is not correct because 7 ones equals 7, and the 7 in 579 is in the tens place.

● 7 tens

This answer is correct because 7 tens equals 70, and the 7 in 579 is in the tens place.

Ⓓ 7 hundreds

This answer is not correct because 7 hundreds equals 700, and the 7 in 579 is in the tens place.

PART FOUR: Learn More About Number Sense

You use number sense when you count.

- When you count, you tell what number comes before or after another number.
- To tell where something is in a row, line, list, or other group, use the number words *first, second, third, fourth, fifth,* and so on.
- To find the total number of items in two or more groups, count each group alone, starting with the number 1. Then add the groups together.

Theo counted the cars in his toy car collection. He used different ways to count the cars. Do Numbers 7 through 10.

7. Theo counted 38 cars. What number comes after 38?

Ⓐ 37 Ⓒ 39
Ⓑ 40 Ⓓ 36

8. Theo has 6 sports cars in his collection. In what place is the sports car that is shaded?

Start

Ⓐ first Ⓒ third
Ⓑ second Ⓓ fourth

9. Theo counted 12 red cars in his car collection. What number comes before 12?

Ⓐ 13 Ⓒ 10
Ⓑ 11 Ⓓ 9

10. The drawing shows how many cars Theo has in two boxes. What problem does the drawing show?

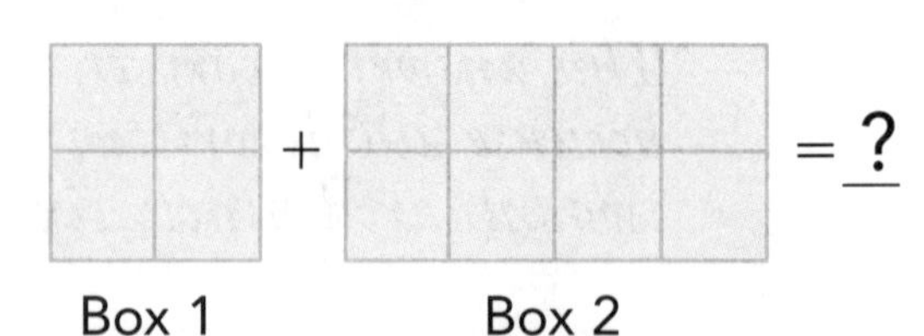

Ⓐ $5 + 4 = 9$ Ⓒ $3 + 7 = 10$
Ⓑ $4 + 8 = 12$ Ⓓ $6 + 2 = 8$

Read what Theo wrote about his visit to the aquarium. Then do Numbers 11 through 14.

The large tank in the aquarium has many different kinds of sea animals.

Here is a list of the different kinds of sea animals in the tank. The list also shows how many of each kind are in the tank.

Sea Animals in the Tank	How Many
Sharks	7
Sea Turtles	4
Sting Rays	11
Sea Snakes	15
Snails, Starfish, Crabs	101

11. Theo saw 15 sea snakes in the tank. What number comes before 15?

Ⓐ 16
Ⓑ 14
Ⓒ 17
Ⓓ 13

12. What animal does Theo list second in his report?

Ⓐ sting rays
Ⓑ sea snakes
Ⓒ sea turtles
Ⓓ sharks

13. Theo counted the snails, starfish, and crabs in the tank. He counted 101. What number is 2 more than 101?

Ⓐ 102
Ⓑ 103
Ⓒ 109
Ⓓ 110

14. What problem shows how many sea turtles and sting rays were in the tank?

Ⓐ $4 + 7 = 11$
Ⓑ $11 + 15 = 26$
Ⓒ $4 + 11 = 15$
Ⓓ $11 + 7 = 18$

PART FIVE: Prepare for a Test

- A test question about number sense may ask for the place value of a digit in a number.
- A test question about number sense may ask you to know a number in its different forms.
- A test question about number sense may ask where something is in a row, line, list, or other group.
- A test question about number sense may ask for the total number of items in two or more groups.

Theo learned some facts about the earth's great rivers. Read the facts. Then do Numbers 15 and 16.

The Earth's Great Rivers

The Nile River, the Amazon River, the Yellow River, and the Mississippi River are some of the earth's greatest rivers. The Nile River is the world's longest river. The Amazon River has more water than the other three rivers added together.

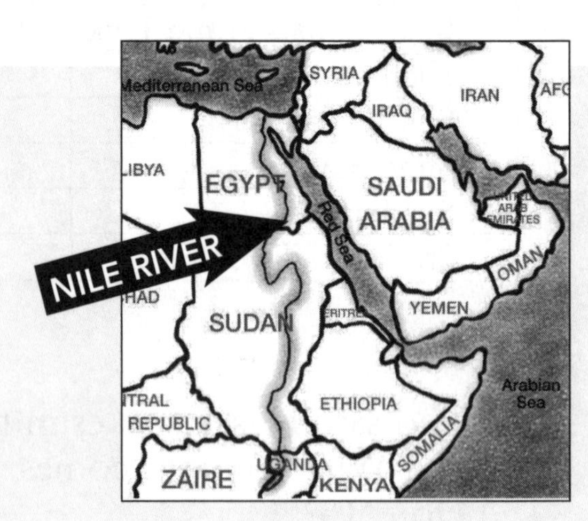

Building Number Sense

15. The Nile River flows through Egypt. It is 4,180 miles long.

What is the place value of the 8 in 4,180?

Ⓐ 4 ones Ⓒ 8 tens
Ⓑ 8 ones Ⓓ 8 hundreds

Building Number Sense

16. In what place is the book that Theo read about the Amazon River?

Start

Ⓐ first Ⓒ third
Ⓑ second Ⓓ fourth

Theo's class watched a movie about storms. Read what Theo did after watching the movie. Then do Numbers 17 and 18.

Twisters

Theo and his classmates watched a movie about different kinds of storms. Then Theo and two other students worked together to make a poster about tornadoes, or twisters. They drew a picture to show the shape of a twister. They also listed many facts about twisters.

Building Number Sense

17. Theo found out that winds inside a twister can reach two hundred fifty miles per hour.

What number should Theo write on the poster to show the speed of winds inside a twister?

Ⓐ 2,150

Ⓑ 205

Ⓒ 250

Ⓓ 25

Building Number Sense

18. Theo made this drawing to show how many twisters his state had in one summer.

What problem does the drawing show?

\+ \+ = ?

June July August

Ⓐ $3 + 4 + 9 = 16$

Ⓑ $5 + 7 + 4 = 16$

Ⓒ $4 + 6 + 2 = 12$

Ⓓ $6 + 6 + 4 = 16$

Strategy Two

USING ESTIMATION

PART ONE: Think About Estimation

WHAT DO YOU KNOW ABOUT ESTIMATION?

You use estimation to find the nearest ten of a number.

Count up or count back from a number to find its nearest ten.

▶ Study the chart and the reason why each estimation to the nearest 10 is correct.

	Number	Nearest Ten
1	21	20
2	25	30
3	28	30

1. The nearest ten of 21 is 20, because 21 is closer to 20 than to 30.

2. The nearest ten of 25 is 30, because 25 is half of the way between 20 and 30. Numbers half of the way between two tens are rounded up.

3. The nearest ten of 28 is 30, because 28 is closer to 30 than to 20.

Estimate each number to the nearest 10. Write the estimate.

a. 46 → ______

b. 32 → ______

c. 18 → ______

d. 24 → ______

e. 35 → ______

f. 52 → ______

g. 55 → ______

h. 61 → ______

You just reviewed how to estimate a number to the nearest 10.

WHAT MORE DO YOU KNOW ABOUT ESTIMATION?

You use estimation to find the nearest hundred of a number.

Count up or count back from a number to find its nearest hundred.

▶ Study the chart and the reason why each estimation to the nearest 100 is correct.

	Number	Nearest Hundred
1	121	100
2	150	200
3	186	200

1. The nearest hundred of 121 is 100, because 121 is closer to 100 than to 200.
2. The nearest hundred of 150 is 200, because 150 is half of the way between 100 and 200. Numbers half of the way between two hundreds are rounded up.
3. The nearest hundred of 186 is 200, because 186 is closer to 200 than to 100.

Estimate each number to the nearest 100. Write the estimate.

a. 104 → ______ e. 498 → ______

b. 248 → ______ f. 543 → ______

c. 384 → ______ g. 612 → ______

d. 410 → ______ h. 675 → ______

You just reviewed how to estimate a number to the nearest 100.

Alone, write six two-digit numbers and six three-digit numbers. Exchange papers with your partner. Find the nearest ten for each two-digit number and the nearest hundred for each three-digit number. Working together, discuss all answers.

PART TWO: Learn About Estimation

Study the estimation chart that Lin's teacher made. As you study, think about each number and its nearest ten.

Number	Nearest Ten
12	10
17	20

The number 12 is between 10 and 20.

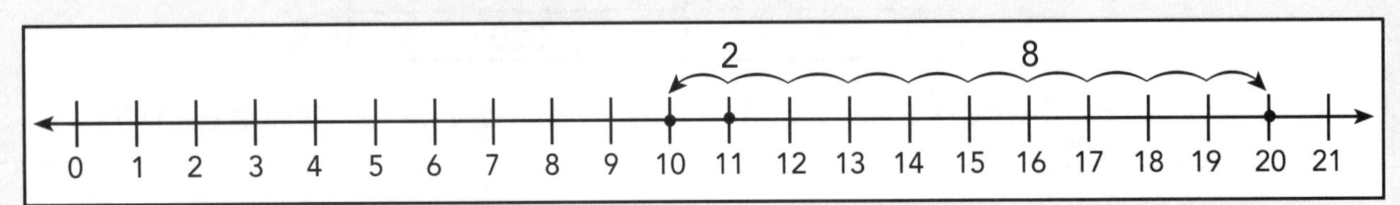

The number 12 is only 2 more than 10. The number 12 is 8 less than 20. Because 12 is closer to 10 than it is to 20, the nearest ten of 12 is 10.

The number 17 is 7 more than 10. The number 17 is only 3 less than 20. Because 17 is closer to 20 than it is to 10, the nearest ten of 17 is 20.

You can find the nearest ten or the nearest hundred of a number.

Number	Nearest Ten	Nearest Hundred
272	270	300
637	640	600

You use **estimation** to find the nearest ten or the nearest hundred of a number.

- Count up and count back from a number to find its nearest ten. Choose the ten that is closer to the number.
- Count up and count back from a number to find its nearest hundred. Choose the hundred that is closer to the number.

Lin made a chart that shows the number of minutes she spent reading each day. Study Lin's chart. Use estimation to figure out about how many minutes Lin read each day. Then do Numbers 1 through 4.

	Monday	Tuesday	Wednesday	Thursday	Friday	TOTAL
Minutes	29	51	46	42	17	185

1. Lin read for 29 minutes on Monday. What is the nearest ten of 29?

Ⓐ 10

Ⓑ 20

Ⓒ 30

Ⓓ 40

2. On Tuesday and Wednesday, Lin read a total of 97 minutes. What is the nearest hundred of 97?

Ⓐ 100

Ⓑ 200

Ⓒ 300

Ⓓ 90

3. Lin read for 51 minutes on Tuesday. About how many minutes did she read?

Ⓐ 30 minutes

Ⓑ 40 minutes

Ⓒ 60 minutes

Ⓓ 50 minutes

4. Lin read for a total of 185 minutes during the week. About how many minutes did she read?

Ⓐ 100 minutes

Ⓑ 200 minutes

Ⓒ 300 minutes

Ⓓ 400 minutes

Talk about your answers to questions 1–4.
Tell why you chose the answers you did.

PART THREE: Check Your Understanding

Remember: You use estimation to find the nearest ten or the nearest hundred of a number.

- Count up and count back from a number to find its nearest ten. Choose the ten that is closer to the number.
- Count up and count back from a number to find its nearest hundred. Choose the hundred that is closer to the number.

Solve this problem. As you work, ask yourself, "What is the nearest ten of the number?"

5. Lin has 28 pencils in a box. About how many pencils are in the box?

Ⓐ 10 pencils
Ⓑ 20 pencils
Ⓒ 30 pencils
Ⓓ 40 pencils

Solve another problem. As you work, ask yourself, "What is the nearest hundred of the number?"

6. Today, 322 tickets for the play at Lin's school were sold. About how many tickets were sold?

Ⓐ 200 tickets
Ⓑ 300 tickets
Ⓒ 400 tickets
Ⓓ 500 tickets

Look at the answer choices for each question.
Read why each answer choice is correct or not correct.

5. Lin has 28 pencils in a box. About how many pencils are in the box?

Ⓐ 10 pencils

This answer is not correct because 10 is 18 less than 28, so 10 is not the nearest ten.

Ⓑ 20 pencils

This answer is not correct because 20 is 8 less than 28, so 20 is not the nearest ten.

● 30 pencils

This answer is correct because 30 is only 2 more than 28, so 30 is the nearest ten.

Ⓓ 40 pencils

This answer is not correct because 40 is 12 more than 28, so 40 is not the nearest ten.

6. Today, 322 tickets for the play at Lin's school were sold. About how many tickets were sold?

Ⓐ 200 tickets

This answer is not correct because 200 is 122 less than 322, so 200 is not the nearest hundred.

● 300 tickets

This answer is correct because 300 is only 22 less than 322, so 300 is the nearest hundred.

Ⓒ 400 tickets

This answer is not correct because 400 is 78 more than 322, so 400 is not the nearest hundred.

Ⓓ 500 tickets

This answer is not correct because 500 is 178 more than 322, so 500 is not the nearest hundred.

PART FOUR: Learn More About Estimation

You can use estimation to check if an answer makes sense. When you add numbers together, you can use estimation to check your sum.

▶ Use estimation to find the nearest ten of each number that you are adding.

Add your estimates together. This total is an estimation of the sum.

Compare the estimated total and the sum. If the two numbers are about the same, the sum is probably correct.

Lin and her sister and brother are cleaning their rooms. Lin wants to make some estimates about the things in the rooms. Do Numbers 7 through 10.

7. Lin picked up 7 socks. Her brother picked up 16 socks. Her sister picked up 12 socks. About how many socks did the children pick up in all?

   Ⓐ 10 socks

   Ⓑ 20 socks

   Ⓒ 30 socks

   Ⓓ 40 socks

8. Lin put 22 books on a shelf and 39 books in a box. To the nearest ten, about how many books did Lin put away?

   Ⓐ 30 books

   Ⓑ 40 books

   Ⓒ 50 books

   Ⓓ 60 books

9. Lin found 88 blocks under the bed. Her sister found 91 more blocks in the closet. To the nearest hundred, about how many blocks did the girls find in all?

   Ⓐ 100 blocks

   Ⓑ 200 blocks

   Ⓒ 300 blocks

   Ⓓ 400 blocks

10. Lin folded 9 shirts. Her brother folded 24 shirts. About how many shirts did the children fold?

   Ⓐ 10 shirts

   Ⓑ 20 shirts

   Ⓒ 30 shirts

   Ⓓ 40 shirts

Read this part of the report that Lin wrote about the space shuttle. Then do Numbers 11 through 14.

The first shuttle went into space in April 1981. Since then, many shuttles have lifted off. The men and women on the shuttle have many jobs to do. They put some machines into orbit. They fix other machines that are already in space. They also do science experiments.

Read some facts about space shuttles in 1998.

Date of Lift-off	Days in Space
January 22, 1998	9 days
April 17, 1998	16 days
June 2, 1998	10 days
October 29, 1998	9 days
December 4, 1998	12 days

11. A shuttle was in space for 16 days in April, 1998. What is the nearest ten of 16?

Ⓐ 10

Ⓑ 20

Ⓒ 30

Ⓓ 40

12. About how long was a shuttle in space in December 1998?

Ⓐ 0 days

Ⓑ 10 days

Ⓒ 20 days

Ⓓ 30 days

13. For about how many days was a shuttle in space during January and October of 1998?

Ⓐ 10 days

Ⓑ 20 days

Ⓒ 30 days

Ⓓ 40 days

14. For about how many days was a shuttle in space during 1998?

Ⓐ 40 days

Ⓑ 50 days

Ⓒ 60 days

Ⓓ 70 days

- A test question about estimation may ask for the nearest ten of a number.
- A test question about estimation may ask for the nearest hundred of a number.
- A test question about estimation may ask for an estimated sum.

Read the facts that Lin learned about animals.
Then do Numbers 15 and 16.

Lin Learns About Animals
Lin learned about some of the fastest animals on earth. She learned that the fastest reptile is a kind of sea turtle. She found out that the fastest land animal is the cheetah.

Using Estimation

15. Lin's teacher told the class that one kind of large sea turtle moves 33 feet every second.

What is the nearest ten of 33?

Ⓐ 3
Ⓑ 20
Ⓒ 30
Ⓓ 40

Using Estimation

16. Lin was surprised to learn that a cheetah can run 92 feet in one second.

About how many feet can a cheetah run in one second?

Ⓐ 70 feet
Ⓑ 80 feet
Ⓒ 90 feet
Ⓓ 100 feet

Lin's class visited the science museum. They learned about the planets. Read what Lin learned at the museum. Then do Numbers 17 and 18.

Spinning Planets

At the museum, Lin's class saw a model of the planets in space. A sign told about each planet. Lin learned that the planets spin while they travel around the sun. She learned that the earth takes 24 hours to spin around once.

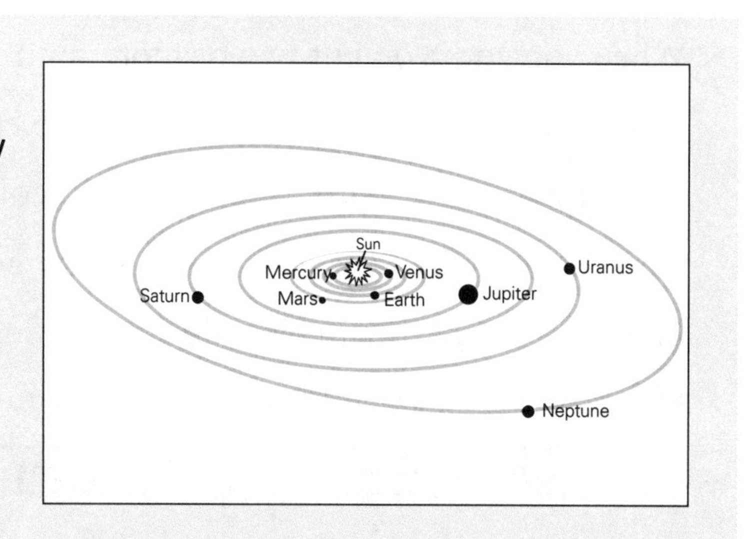

Using Estimation

17. Lin learned that Venus takes 243 days to spin around once.

What is the nearest hundred of 243?

Ⓐ 100

Ⓑ 200

Ⓒ 300

Ⓓ 240

Using Estimation

18. Lin estimated the total number of moons of Jupiter, Neptune, and Saturn.

About how many moons do the three planets have?

Planet	Number of Moons
Jupiter	16
Neptune	8
Saturn	18

Ⓐ 20 moons

Ⓑ 30 moons

Ⓒ 40 moons

Ⓓ 50 moons

Strategy Three

APPLYING ADDITION

PART ONE: Think About Addition

WHAT DO YOU KNOW ABOUT ADDITION?

When you add, you put two or more numbers together to find the total, or sum.

The numbers that you add together are called addends.

Addends may be written in rows or in columns.

Addends may be placed in any order to get the same total.

Count on from one of the addends to find the sum.

▶ Study the chart and number lines to find each sum.

Use any order.	Count on. Use a number line.
$4 + 5 = 9$	0 1 2 3 4 5 6 7 8 9 10
$5 + 4 = 9$	0 1 2 3 4 5 6 7 8 9 10

Count on from the first addend to find the sum for each problem.

a. $2 + 3 =$ ______

$3 + 2 =$ ______

b. $4 + 3 =$ ______

$3 + 4 =$ ______

c. $6 + 4 =$ ______

$4 + 6 =$ ______

d. $7 + 2 =$ ______

$2 + 7 =$ ______

e. $6 + 5 =$ ______

$5 + 6 =$ ______

f. $5 + 8 =$ ______

$8 + 5 =$ ______

You just reviewed counting on from one addend to arrive at a sum.

WHAT MORE DO YOU KNOW ABOUT ADDITION?

When an addend has two digits, think about the place value of each digit.

Think about the value of the digit in the ones column.

Think about the value of the digit in the tens column.

When you add, line up the ones and the tens in columns.

▶ Study the problem 16 + 25 to find the sum.
Note that 6 ones + 5 ones are 11 ones.
Regroup 10 of the ones as 1 ten.
Add the tens to get 4 tens.
The sum is 40 + 1 or 41.

tens	ones
$\overset{1}{1}$	6
+ 2	5
4	1

Arrange the addends in columns. Then, find the sum of each addition problem.

Problems	**Problems in Columns**
a. 32 + 24 = ________	a.
b. 41 + 38 = ________	b.
c. 33 + 18 = ________	c.
d. 36 + 45 = ________	d.
e. 48 + 17 = ________	e.

You just reviewed how to arrange two-digit addends in columns before finding the sum.

Together, write four addition problems. Let each addend in the problems have two digits. Arrange the problems in columns. Alone, solve two problems while your partner solves the other two problems. When finished, work together to discuss the results.

PART TWO: Learn About Addition

Study the problem that Hector's teacher wrote on the chalkboard.
As you study, think about the three ways that Hector solved the problem.

Problem: A boy has 5 toy airplanes.
His aunt gave him 4 more toy airplanes.
How many toy airplanes does the boy now have?

Three Ways That Hector Solves the Problem		
1. I write the numbers across. I use boxes to help me count.	$5 + 4 = 9$	❑❑❑❑❑ + ❑❑❑❑
2. I use a different order. I use boxes to help me count.	$4 + 5 = 9$	❑❑❑❑ + ❑❑❑❑❑
3. I use a number line to count on from the number 5.	$5 + 4 = 9$	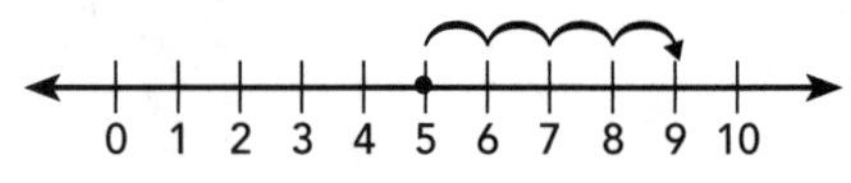

Answer: The boy has 9 toy airplanes.

When you add, you put two or more numbers together to find the total, or sum. The numbers that you add together are called addends.

Addends may be written in rows or in columns.
Addends may be put in any order. The total of the addends does not change.

$2 + 6 = 8$ $6 + 2 = 8$ $\begin{array}{r} 6 \\ +2 \\ \hline 8 \end{array}$ $\begin{array}{r} 2 \\ +6 \\ \hline 8 \end{array}$

You use **addition** to find the total of two or more addends.

- Write addends in any order, in rows or in columns, and get the same total.
- Count on from one of the addends to find the total.

Study another one of Hector's problems. Look at the ways that Hector solved the problem. Then do Numbers 1 through 4.

Problem: Mrs. Lee has 4 apples in a bowl.
She has 3 more apples in a bag.
How many apples does Mrs. Lee have all together?

Use any order.	Count on. Use a number line.
4 + 3 = 7	0 1 2 3 4 5 6 7 8 9 10
3 + 4 = 7	0 1 2 3 4 5 6 7 8 9 10

Answer: Mrs. Lee has 7 apples.

1. Hector has 2 crayons on his desk.
He has 3 crayons in his hand.
How many crayons does he have all together?

Ⓐ 5 crayons
Ⓑ 6 crayons
Ⓒ 4 crayons
Ⓓ 7 crayons

2. Hector's mom planted 9 trees.
Then she planted 1 more tree.
How many trees did she plant in all?

Ⓐ 12 trees
Ⓑ 10 trees
Ⓒ 13 trees
Ⓓ 15 trees

3. This morning, Hector saw 4 robins.
This afternoon, he saw 2 robins.
What is the total number of robins that Hector saw today?

Ⓐ 5 robins
Ⓑ 9 robins
Ⓒ 6 robins
Ⓓ 8 robins

4. Hector has 7 aunts and 2 uncles.
How many aunts and uncles does he have all together?

Ⓐ 10 aunts and uncles
Ⓑ 8 aunts and uncles
Ⓒ 11 aunts and uncles
Ⓓ 9 aunts and uncles

Talk about your answers to questions 1–4.
Tell why you chose the answers you did.

PART THREE: Check Your Understanding

Remember: You use addition to find the total of two or more addends.

- Write addends in any order, in rows or in columns, and get the same total.
- Count on from one of the addends to find the total.

Solve this problem. As you work, ask yourself, "How can I use the order of addends to find their total?"

5. This month, Hector stayed home 5 days because he was sick. Last month, Hector stayed home 3 days because he was sick. What is the total number of days that he stayed home?

Ⓐ 7 days

Ⓑ 6 days

Ⓒ 8 days

Ⓓ 2 days

Solve another problem. As you work, ask yourself, "How can I use counting on to help me find the total?"

6. Hector made 6 snowballs, and his sister Elena made 5 snowballs. How many snowballs did the children make together?

Ⓐ 13 snowballs

Ⓑ 1 snowball

Ⓒ 9 snowballs

Ⓓ 11 snowballs

Look at the answer choices for each question.
Read why each answer choice is correct or not correct.

5. This month, Hector stayed home 5 days because he was sick. Last month, Hector stayed home 3 days because he was sick. What is the total number of days that he stayed home?

Ⓐ 7 days

This answer is not correct because $5 + 3 = 8$ and $3 + 5 = 8$, not 7.

Ⓑ 6 days

This answer is not correct because $5 + 3 = 8$ and $3 + 5 = 8$, not 6.

● 8 days

This answer is correct because $5 + 3 = 8$ and $3 + 5 = 8$.

Ⓓ 2 days

This answer is not correct because $5 + 3 = 8$ and $3 + 5 = 8$, not 2.

6. Hector made 6 snowballs, and his sister Elena made 5 snowballs. How many snowballs did the children make together?

Ⓐ 13 snowballs

This answer is not correct because counting on 5 more from 6 will get to 11, not 13.

Ⓑ 1 snowball

This answer is not correct because counting on 5 more from 6 will get to 11, not 1.

Ⓒ 9 snowballs

This answer is not correct because counting on 5 more from 6 will get to 11, not 9.

● 11 snowballs

This answer is correct because counting on 5 more from 6 will get to 11.

PART FOUR: Learn More About Addition

You use addition to find the sum, or total, of addends. When an addend has two digits, think about the value of the digit in the ones place and the value of the digit in the tens place.

▶ When you write the addends, line up the ones and tens columns.

Add the ones column, and write the sum. If the sum of the ones is more than 10, regroup 10 of the ones as 1 ten. Write 1 in the tens column to stand for the 10 ones you regrouped.

Add the tens column.

tens	ones
1	
2	6
+ 3	8
6	4

Hector's family is moving. Hector helped pack.
Do Numbers 7 through 10.

7. Hector put 12 books into one box and 11 books into another box. How many books did he put into the boxes?

Ⓐ 19 books

Ⓑ 23 books

Ⓒ 13 books

Ⓓ 33 books

8. In one box, Hector's mom packed 4 red towels, 5 green towels, and 12 white towels. How many towels did she pack?

Ⓐ 21 towels

Ⓑ 16 towels

Ⓒ 28 towels

Ⓓ 14 towels

9. Hector put 24 cups into a box. His dad put 18 cups into another box. How many cups did Hector and his dad pack?

Ⓐ 42 cups

Ⓑ 26 cups

Ⓒ 32 cups

Ⓓ 41 cups

10. Hector counted all the packed boxes. There were 29 large boxes and 17 small boxes. What was the total number of boxes?

Ⓐ 30 boxes

Ⓑ 36 boxes

Ⓒ 46 boxes

Ⓓ 29 boxes

Read the postcard message that Hector wrote to his friend Maria. Then do Numbers 11 through 14.

Dear Maria,

Our new house is great. We have a big yard. Our dog Jump plays in the park at the end of our street. All the children from the neighborhood play there too.

Please write soon.

Your friend,
Hector

11. Hector counted the houses on his street. On one side, he counted 15 houses. On the other side, he counted 16. What is the total number of houses on the street?

Ⓐ 21 houses

Ⓑ 31 houses

Ⓒ 18 houses

Ⓓ 35 houses

12. Jump likes to bury leaves in the yard. He buried 34 leaves in one hole. He buried 22 leaves in another hole. How many leaves did Jump bury?

Ⓐ 66 leaves

Ⓑ 49 leaves

Ⓒ 44 leaves

Ⓓ 56 leaves

13. On Monday, Hector met 2 boys at the park. On Tuesday, he met 4 girls. On Wednesday, he met 8 more boys and girls. How many children did he meet in all?

Ⓐ 14 children

Ⓑ 24 children

Ⓒ 17 children

Ⓓ 20 children

14. Hector sent Maria 1 birthday card, 15 postcards, and 8 letters. How many cards and letters did Hector send Maria all together?

Ⓐ 28 cards and letters

Ⓑ 34 cards and letters

Ⓒ 24 cards and letters

Ⓓ 22 cards and letters

- A test question about addition may ask for the sum of two or more addends.
- A test question about addition may ask for the sum of numbers with two digits.

Read about Hector and his grandmother.
Then do Numbers 15 and 16.

Hector Helps His Grandmother

Hector's grandmother owns a restaurant. Every morning, she drives to the open market. Sometimes, she takes Hector along. They pick out fresh food to use in the restaurant.

Applying Addition

15. Hector counted the number of fruit carts in three lanes.

One lane had 8 carts.
One lane had 7 carts.
One lane had 3 carts.

How many carts did Hector count?

Ⓐ 26 carts
Ⓑ 17 carts
Ⓒ 18 carts
Ⓓ 28 carts

Applying Addition

16. Hector's grandmother bought 15 pounds of large tomatoes. She bought 24 pounds of cherry tomatoes.

How many pounds of tomatoes did she buy all together?

Ⓐ 32 pounds
Ⓑ 39 pounds
Ⓒ 29 pounds
Ⓓ 48 pounds

Read some facts that Hector learned about vegetables. Then do Numbers 17 and 18.

Learn Your Vegetables

Carrots and beets come from roots. Peas are vegetables that come from seeds. Spinach and lettuce come from leaves. Peppers, cucumbers, and squash are from flowering plants.

Applying Addition

17. Hector and his mom planted four rows of lettuce plants and four rows of spinach plants.

Lettuce		Spinach
5	+	8
6	+	7
9	+	4
3	+	10

How many plants were in each row?

Ⓐ 15 plants

Ⓑ 23 plants

Ⓒ 11 plants

Ⓓ 13 plants

Applying Addition

18. The 3 pepper plants in Hector's garden were covered with small white flowers.

One plant had 14 flowers.
One plant had 21 flowers.
One plant had 17 flowers.

What was the total number of flowers on the three plants?

Ⓐ 52 flowers

Ⓑ 62 flowers

Ⓒ 42 flowers

Ⓓ 45 flowers

Strategies One–Three REVIEW

PART ONE: Read a Story

Read the story about Sophie and Joe. Then do Numbers 1 through 6.

Number Crunching

"Hi. My name is Sophie. My partner's name is Joe. We're number detectives. Our job is to find numbers. We find numbers everywhere. When we find them, we write them down. Later, we use them to make up problems."

"Yesterday, Joe and I searched our neighborhood. We were pretty lucky. We found a lot of numbers. We wrote them down in this chart."

PLACE	NUMBER
Mr. Van's Gas Station	
Customers in one hour	18
Gas pumped	419 gallons
Main Street	
Speed limit sign	25
House number	109
Sacarro's Pizza Palace	
Customers in one hour	37
Cheese pizzas sold	13
Pepperoni pizzas sold	9
The Farm Stand	
Customers in one hour	24
Pounds of grapes sold	9
Pounds of apples sold	7
Pounds of plums sold	4

Building Number Sense

1. Which of these shows how the number 419 can be written?

Ⓐ forty-nine
Ⓑ four hundred nineteen
Ⓒ four hundred nine
Ⓓ forty

Building Number Sense

2. What house number on Main Street comes after 109?

Ⓐ 100
Ⓑ 101
Ⓒ 110
Ⓓ 108

Using Estimation

3. Sacarro's Pizza Palace had 37 customers in one hour.

Which of these is the nearest ten of 37?

Ⓐ 3
Ⓑ 20
Ⓒ 30
Ⓓ 40

Using Estimation

4. About how many customers did the gas station, pizza place, and farm stand have in one hour?

Ⓐ 70 customers
Ⓑ 80 customers
Ⓒ 90 customers
Ⓓ 100 customers

Applying Addition

5. What was the total number of cheese and pepperoni pizzas sold?

Ⓐ 103 pizzas
Ⓑ 12 pizzas
Ⓒ 23 pizzas
Ⓓ 22 pizzas

Applying Addition

6. How many pounds of fruit were sold at the farm stand?

Ⓐ 44 pounds
Ⓑ 10 pounds
Ⓒ 20 pounds
Ⓓ 18 pounds

PART TWO: Read an Article

Read the questions and answers found in the article about frogs. Then do Numbers 7 through 12.

Where do you see frogs?

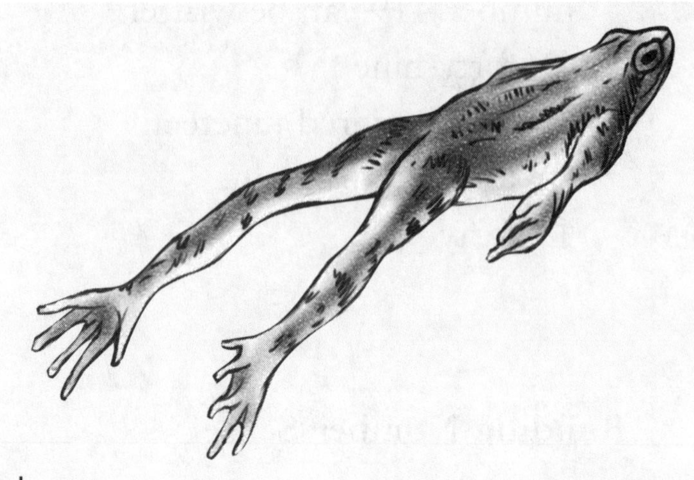

You might see frogs on lily pads in a pond or on a log in the woods. You might see frogs hopping through the grass in your own backyard.

Frogs live in other places, too. The water-holding frog lives in the desert. Some frogs live in trees. The United States is home to about 25 kinds of tree frogs. About 300 kinds of tree frogs live in tropical places.

Frogs have been around for a very long time. Scientists believe that frogs were walking, or maybe jumping, around the earth 190 million years ago. Those frogs had to be careful. Some very large dinosaurs were also walking around!

Today, there are more than 3,000 kinds of frogs and toads. Many are green. Some are brown. Some have red eyes. Some have yellow, blue, orange, or purple spots. Some frogs are smaller than 1 inch. Some are 12 inches long.

Most kinds of frogs lay their eggs in water in early spring. It can take from 3 to 25 days for baby frogs, called tadpoles, to hatch. Within a few weeks, tadpoles lose their tails, grow legs, and become frogs.

What is the difference between frogs and toads?

Toads are a kind of frog. Toads usually have thick, dry, warty skin. They spend less time in the water than frogs do. Frogs usually have smooth, slimy skin. They're great swimmers.

Toads have shorter hind legs than frogs. Because of this, toads are not great jumpers. On the other hand, a frog has long, strong hind legs that are perfect for jumping. A frog usually can jump 20 times its body length. In a frog contest in South Africa, a frog jumped 33 feet 5.5 inches, or about 1,020 cm. That was a record leap!

Building Number Sense

7. A frog in South Africa jumped about 1,020 cm.

What is the place value of the 2 in 1,020?

Ⓐ 2 thousands
Ⓑ 2 ones
Ⓒ 2 tens
Ⓓ 2 hundreds

Using Estimation

10. In the United States and tropical places together, there are about 325 different kinds of tree frogs.

Which of these is the nearest hundred of 325?

Ⓐ 200
Ⓑ 300
Ⓒ 350
Ⓓ 400

Building Number Sense

8. Look at the lily pads.

In what position is the lily pad without a frog?

Ⓐ first
Ⓑ second
Ⓒ third
Ⓓ fourth

Applying Addition

11. How many frogs are in each place?

Place	Frogs
lily pads	6 + 8
log	7 + 7
tree	5 + 9
grass	4 + 10

Ⓐ 14 frogs
Ⓑ 12 frogs
Ⓒ 21 frogs
Ⓓ 16 frogs

Using Estimation

9. A frog ate 96 mosquitoes in one night.

About how many mosquitoes did the frog eat?

Ⓐ 70 mosquitoes
Ⓑ 80 mosquitoes
Ⓒ 90 mosquitoes
Ⓓ 100 mosquitoes

Applying Addition

12. One tadpole changed into a frog in 26 days. A second tadpole took 11 more days.

How long did it take the second tadpole to change into a frog?

Ⓐ 29 days
Ⓑ 37 days
Ⓒ 35 days
Ⓓ 28 days

Strategy Four APPLYING SUBTRACTION

PART ONE: Think About Subtraction

WHAT DO YOU KNOW ABOUT SUBTRACTION?

When you subtract, you find the difference between two numbers.

To subtract, arrange the two numbers in order.

Put the larger number at the top.

Subtract the smaller number.

The answer in subtraction is the difference.

▶ Arrange each subtraction problem in columns.
Subtract the smaller number from the larger to get the difference.
The first problem has been completed for you.

Problems	**Problems in Columns**
a. $6 - 4 =$ 2	a. $\begin{array}{r} 6 \\ -\ 4 \\ \hline 2 \end{array}$
b. $9 - 5 =$ _____	b.
c. $8 - 2 =$ _____	c.
d. $10 - 5 =$ _____	d.
e. $12 - 4 =$ _____	e.
f. $16 - 7 =$ _____	f.
g. $20 - 11 =$ _____	g.
h. $22 - 12 =$ _____	h.
i. $25 - 13 =$ _____	i.

You just reviewed finding the difference between two numbers.

WHAT DO YOU KNOW ABOUT REGROUPING IN SUBTRACTION?

When the numbers in subtraction have two digits, think about the place value of each digit.

Think about the value of the digit in the ones column.

Think about the value of the digit in the tens column.

When you subtract, line up the ones and the tens in the columns.

▶ Arrange each subtraction problem in columns.
Subtract the smaller number from the larger number to get the difference. You may need to regroup 1 ten as 10 ones.
The first problem has been completed for you.

Problems	**Problems in Columns**
a. 24 − 16 = <u>8</u>	a. $\begin{array}{r} 24 \\ -\ 16 \\ \hline 8 \end{array}$
b. 26 − 18 = _____	b.
c. 24 − 15 = _____	c.
d. 35 − 18 = _____	d.
e. 32 − 16 = _____	e.
f. 30 − 12 = _____	f.
g. 31 − 14 = _____	g.
h. 36 − 18 = _____	h.
i. 25 − 13 = _____	i.

You just reviewed subtraction of two-digit numbers with regrouping.

Write four subtraction problems. Let each number in the problem be a two-digit number. Have each problem require regrouping. Alone, solve two of the problems, while your partner solves the other two problems. When finished, work together to discuss each solution.

PART TWO: Learn About Subtraction

Study the problem that Reina's teacher wrote on the chalkboard.
As you study, think about the three ways that Reina solved the problem.

Problem: A puppy buried 9 bones in the yard in one week.
The next week, the puppy dug up 4 bones.
How many bones were still buried?

Three Ways That Reina Solves the Problem	
1. I draw a picture. I cross out 4 boxes. I count the boxes that are left.	
2. I use a number line. I count 4 back from 9 to find the answer.	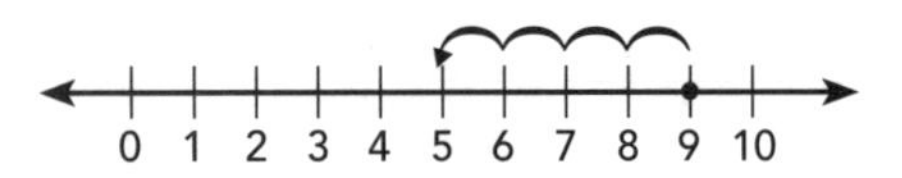
3. I write the numbers across. I write the larger number first. I subtract the smaller number from the larger number.	$9 - 4 = 5$

Answer: There were 5 bones still buried.

When you subtract, you find the difference between two numbers. The difference is always smaller than the number subtracted from, unless you subtract 0.

You can write a subtraction problem in a row. Put the larger number first.
You can write a subtraction problem in a column. Put the larger number on top.

$10 - 2 = 8$

$$\begin{array}{r} 10 \\ -\ 2 \\ \hline 8 \end{array}$$

You use **subtraction** to find the difference between two numbers.

- Write numbers in a subtraction problem in a row or in a column.
- Count back from the larger number to find the difference.

Study another one of Reina's problems. Look at the ways that Reina solved the problem. Then do Numbers 1 through 4.

Problem: A girl read 7 books in April and May.
She read 5 books in April.
How many books did the girl read in May?

Use boxes to count. Take away or cross out.	Count back. Use a number line.
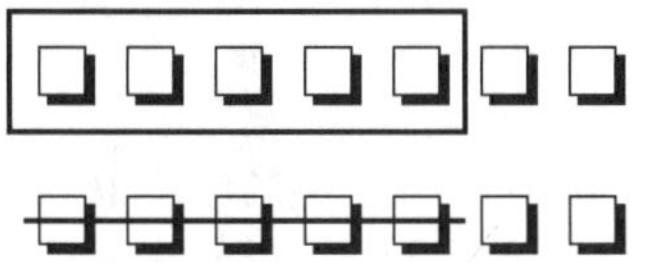 ❑❑❑❑❑❑❑ (5 boxes grouped, 2 boxes apart); 5 boxes crossed out, 2 boxes left	 Number line 0 1 2 3 4 5 6 7 8 9 10, counting back from 7 to 2

Answer: The girl read 2 books in May.

1. Reina's dad brought home 10 bags of groceries. Reina carried 4 bags into the house. How many bags did Reina's dad carry?

   Ⓐ 7 bags
   Ⓑ 5 bags
   Ⓒ 6 bags
   Ⓓ 4 bags

2. Reina took 12 oranges from one bag. She put 8 oranges into the refrigerator. How many oranges did she leave out?

   Ⓐ 2 oranges
   Ⓑ 5 oranges
   Ⓒ 6 oranges
   Ⓓ 4 oranges

3. Reina's dad bought 5 pounds of potatoes. He used 2 pounds of potatoes for dinner. How many pounds of potatoes were left?

   Ⓐ 3 pounds
   Ⓑ 1 pound
   Ⓒ 4 pounds
   Ⓓ 2 pounds

4. Reina peeled 8 carrots. Her mother put 6 of the carrots into the soup and the rest into the salad. How many carrots went into the salad?

   Ⓐ 4 carrots
   Ⓑ 2 carrots
   Ⓒ 1 carrot
   Ⓓ 6 carrots

Talk about your answers to questions 1–4.
Tell why you chose the answers you did.

PART THREE: Check Your Understanding

Remember: You use subtraction to find the difference between two numbers.

- Write numbers in a subtraction problem in a row or in a column.
- Count back from the larger number to find the difference.

Solve this problem. As you work, ask yourself, "What number is the larger number? What number do I take away?"

5. There are 14 girls in Reina's class. On Tuesday, 5 of the girls signed up for the school talent show. How many girls from Reina's class will *not* be in the talent show?

   Ⓐ 8 girls

   Ⓑ 9 girls

   Ⓒ 11 girls

   Ⓓ 6 girls

Solve another problem. As you work, ask yourself, "How can I use counting back to help me find the difference?"

6. Reina looked out into the audience. She counted 11 chairs in the first row. There were people sitting in 4 of the chairs. How many chairs in the first row were empty?

   Ⓐ 3 chairs

   Ⓑ 6 chairs

   Ⓒ 7 chairs

   Ⓓ 5 chairs

Look at the answer choices for each question.
Read why each answer choice is correct or not correct.

5. There are 14 girls in Reina's class. On Tuesday, 5 of the girls signed up for the school talent show. How many girls from Reina's class will *not* be in the talent show?

Ⓐ 8 girls

This answer is not correct because if you take 5 away from 14, you have 9 left, not 8.

● 9 girls

This answer is correct because if you take 5 away from 14, you have 9 left. So, $14 - 5 = 9$.

Ⓒ 11 girls

This answer is not correct because if you take 5 away from 14, you have 9 left, not 11.

Ⓓ 6 girls

This answer is not correct because if you take 5 away from 14, you have 9 left, not 6.

6. Reina looked out into the audience. She counted 11 chairs in the first row. There were people sitting in 4 of the chairs. How many chairs in the first row were empty?

Ⓐ 3 chairs

This answer is not correct because if you count back 4 from 11, you get 7, not 3.

Ⓑ 6 chairs

This answer is not correct because if you count back 4 from 11, you get 7, not 6.

● 7 chairs

This answer is correct because if you count back 4 from 11, you get 7. So, $11 - 4 = 7$.

Ⓓ 5 chairs

This answer is not correct because if you count back 4 from 11, you get 7, not 5.

PART FOUR: Learn More About Subtraction

You use subtraction to find the difference between two numbers. When one or both of the numbers have two digits, think about the value of the digit in the ones place and the value of the digit in the tens place.

▶ When you subtract, line up the ones and tens columns.

First, look at the ones column.

If the number you are taking away is larger than the number you are taking away from, regroup 1 ten as 10 ones. Take away 1 ten from the tens column. Add 10 to the number in the ones column.

Find the difference. Subtract the ones. Then subtract the tens.

tens	ones
2	12
~~3~~	~~2~~
− 1	8
1	4

Reina and her dad built a model plane. Do Numbers 7 through 10.

7. Reina's dad cut 23 short pieces of wood. He used 19 pieces to build the plane. How many pieces of wood did he have left?

Ⓐ 4 pieces

Ⓑ 12 pieces

Ⓒ 6 pieces

Ⓓ 14 pieces

8. There were 36 nails. Reina counted out 17 nails for her dad to use. How many nails were left?

Ⓐ 17 nails Ⓒ 19 nails

Ⓑ 22 nails Ⓓ 25 nails

9. It took 21 hours to build and paint the plane. Reina and her dad spent 16 hours building the plane. How many hours did they spend painting it?

Ⓐ 7 hours Ⓒ 15 hours

Ⓑ 8 hours Ⓓ 5 hours

10. Reina had 30 stickers. She put 14 of the stickers on the plane. How many of the stickers were left in the package?

Ⓐ 15 stickers

Ⓑ 16 stickers

Ⓒ 22 stickers

Ⓓ 24 stickers

Read this story about Reina and her friends. Then do Numbers 11 through 14.

Reina has a collection of model planes. Some of her friends have collections, too. Chan has a collection of coins from different countries. Beth has a collection of seashells. Tomás collects action figures.

11. Reina has 20 model planes in her collection. She and her dad built 11 of them. Her brother built the others. How many of the model planes in Reina's collection did her brother build?

Ⓐ 12 model planes

Ⓑ 7 model planes

Ⓒ 9 model planes

Ⓓ 11 model planes

12. Chan has 31 coins from Spain. He put 18 of them into a coin album. How many coins from Spain did Chan leave out of the coin album?

Ⓐ 13 coins

Ⓑ 19 coins

Ⓒ 23 coins

Ⓓ 20 coins

13. Beth has 45 seashells that she found on beaches in Florida and South Carolina. She found 19 shells on beaches in Florida. How many of Beth's shells are from beaches in South Carolina?

Ⓐ 33 seashells

Ⓑ 27 seashells

Ⓒ 34 seashells

Ⓓ 26 seashells

14. Tomás has 22 action figures. He brought 16 of them to Reina's house to play. How many action figures did he leave at home?

Ⓐ 16 action figures

Ⓑ 6 action figures

Ⓒ 18 action figures

Ⓓ 9 action figures

- A test question about subtraction may ask for the difference between two numbers.
- A test question about subtraction may ask for the difference between numbers with two digits.

Read about Reina's class project. Then do Numbers 15 and 16.

Counting Insects

Ms. Park took all the students in Reina's second-grade class to the pond behind the school to count insects. Pairs of students looked for different kinds of insects to count. Some counted butterflies. Some counted dragonflies. Some counted other insects.

Applying Subtraction

15. Reina and Chan counted 13 butterflies. Reina counted 9 of them.

How many butterflies did Chan count?

Ⓐ 4 butterflies

Ⓑ 2 butterflies

Ⓒ 7 butterflies

Ⓓ 5 butterflies

Applying Subtraction

16. The students counted a total of 54 insects. All together, they counted 28 butterflies and dragonflies.

How many other insects did they count?

Ⓐ 33 insects

Ⓑ 29 insects

Ⓒ 31 insects

Ⓓ 26 insects

Read how Reina and her classmates watched butterflies grow. Then do Numbers 17 and 18.

Growing Butterflies

Ms. Park's class took care of some furry caterpillars. They took notes and drew pictures when the caterpillars changed into beautiful butterflies. There were monarch butterflies and painted lady butterflies. Reina's group took care of the monarch butterflies. The other group took care of the painted lady butterflies. Then the children released all the butterflies near the school pond.

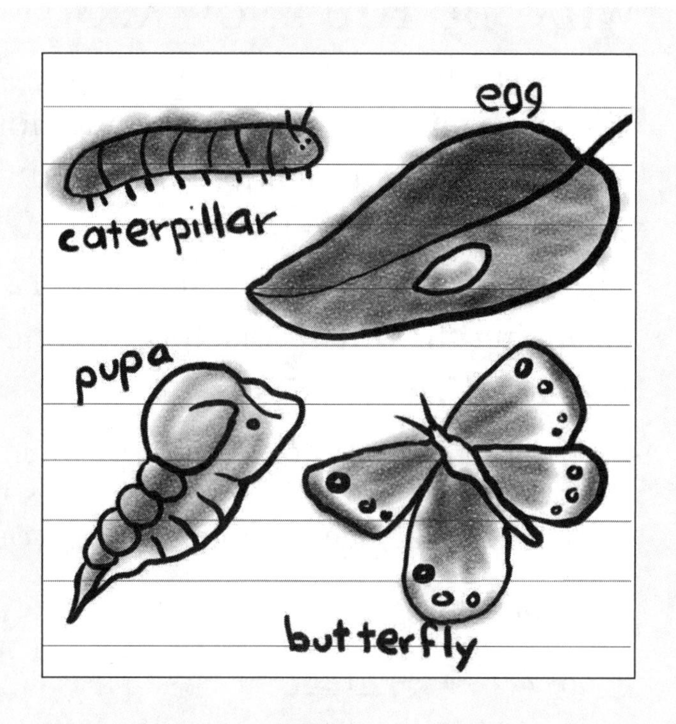

Applying Subtraction

17. There was a total of 35 caterpillars. There were 28 caterpillars that became painted lady butterflies.

How many caterpillars became monarch butterflies?

Ⓐ 17 caterpillars
Ⓑ 13 caterpillars
Ⓒ 7 caterpillars
Ⓓ 5 caterpillars

Applying Subtraction

18. It took 23 days for all the caterpillars to change into butterflies. After 20 days, only two of the monarch caterpillars had *not* changed into butterflies.

How many more days did it take those two monarch caterpillars to become butterflies?

Ⓐ 3 days
Ⓑ 13 days
Ⓒ 15 days
Ⓓ 8 days

Strategy Five APPLYING MULTIPLICATION

PART ONE: Think About Multiplication

WHAT DO YOU KNOW ABOUT MULTIPLICATION?

Multiplication is repeated addition.

The problem 3×4 is also the addition problem $4 + 4 + 4$.

The numbers you multiply are called factors.

The answer is called the product.

The product is 12. The sum is also 12.

Changing the order of factors does not change the product.

$3 \times 4 = 12$ and $4 \times 3 = 12$

▶ Arrange each multiplication problem in columns. Multiply the factors to find the product. Add the addends to show repeated addition. The first problem has been completed for you.

Problems	Problems in Columns
a. $2 \times 6 =$ 12 $6 + 6 =$ 12	a. $\begin{array}{r} 4 \\ \times\ 3 \\ \hline 12 \end{array}$
b. $3 \times 2 =$ _____ $2 + 2 + 2 =$ _____	b.
c. $4 \times 2 =$ _____ $2 + 2 + 2 + 2 =$ _____	c.
d. $2 \times 8 =$ _____ $8 + 8 =$ _____	d.
e. $4 \times 4 =$ _____ $4 + 4 + 4 + 4 =$ _____	e.
f. $2 \times 7 =$ _____ $7 + 7 =$ _____	f.
g. $3 \times 8 =$ _____ $8 + 8 + 8 =$ _____	g.
h. $5 \times 6 =$ _____ $6 + 6 + 6 + 6 + 6 =$ _____	h.

You just reviewed how multiplication is repeated addition.

WHAT DO YOU KNOW ABOUT SKIP-COUNTING IN MULTIPLICATION?

You can skip-count to find a product.

Look at one factor. Count by that number.

Skip-count as many times as the second factor.

▶ Study the problem. Skip-count to find the product.
$6 \times 4 = 24$
Count the number 4, six times.
4, 8, 12, 16, 20, 24
The product of 6×4 is 24.

Complete each multiplication problem. Skip-count to find each product.

a. $3 \times 5 =$ _____
Skip-count the number _____, three times.
Skip-count: ____
The product of 3×5 is _____.

b. $4 \times 3 =$ _____
Skip-count the number _____, four times.
Skip-count: ____
The product of 4×3 is _____.

c. $5 \times 9 =$ _____
Skip-count the number _____, five times.
Skip-count: ____
The product of 5×9 is _____.

d. $4 \times 8 =$ _____
Skip-count the number _____, four times.
Skip-count: ____
The product of 4×8 is _____.

You just used skip-counting to find a product.

Together, write 4 multiplication problems using one-digit factors. Alone, each partner solves two of the problems. Together, discuss how you use skip-counting to find a product.

PART TWO: Learn About Multiplication

Study the problem that Lisa's teacher wrote on the chalkboard.
As you study, think about the five ways that Lisa solved the problem.

Problem: Matt has 3 rows of counters.
There are 4 counters in each row.
How many counters does Matt have in all?

Five Ways That Lisa Solves the Problem	
1. I draw a picture and count.	❑ ❑ ❑ ❑ ❑ ❑ ❑ ❑ ❑ ❑ ❑ ❑
2. I write a repeated addition sentence.	$4 + 4 + 4 = 12$
3. I write a different repeated addition sentence.	$3 + 3 + 3 + 3 = 12$
4. I write a multiplication sentence.	$3 \times 4 = 12$
5. I change the order to write a different multiplication sentence.	$4 \times 3 = 12$

Answer: Matt has 12 counters.

When you multiply, you use repeated addition to find a total.
The numbers that you multiply are called factors. The total is called a product.
You may change the order of the factors. The product does not change.

$2 \times 3 = 6 \qquad 3 \times 2 = 6$

You use **multiplication** to find the product of two factors.

- To help find a product, write a repeated addition sentence for the problem. Then write a multiplication sentence for the same problem.
- Changing the order of the factors does not change the product.

Study another one of Lisa's problems. Look at the ways that Lisa solved the problem. Then do Numbers 1 through 4.

Problem: There were 2 clowns in the parade.
Each clown had 5 balloons.
How many balloons did the clowns have all together?

Write a repeated addition sentence.	$5 + 5 = 10$ OR $2 + 2 + 2 + 2 + 2 = 10$
Write a multiplication sentence.	$2 \times 5 = 10$ OR $5 \times 2 = 10$

Answer: The clowns had 10 balloons.

1. Lisa saw 5 rows of horses. There were 4 horses in each row. How many horses were there?

Ⓐ 40 horses

Ⓑ 9 horses

Ⓒ 15 horses

Ⓓ 20 horses

2. There were 6 elephants in the parade. Lisa figured out the total number of legs on the elephants. How many legs were there?

Ⓐ 20 legs

Ⓑ 24 legs

Ⓒ 18 legs

Ⓓ 32 legs

3. Lisa counted 5 marching bands. Each band had 6 drummers. What was the total number of drummers?

Ⓐ 30 drummers

Ⓑ 20 drummers

Ⓒ 25 drummers

Ⓓ 50 drummers

4. Lisa saw 3 fire trucks at the end of the parade. There were 7 firefighters riding on each fire truck. How many firefighters were there in all?

Ⓐ 23 firefighters

Ⓑ 24 firefighters

Ⓒ 18 firefighters

Ⓓ 21 firefighters

Talk about your answers to questions 1–4.
Tell why you chose the answers you did.

PART THREE: Check Your Understanding

Remember: You use multiplication to find the product of two factors.

- To help find a product, write a repeated addition sentence for the problem. Then write a multiplication sentence for the same problem.
- Changing the order of the factors does not change the product.

Solve this problem. As you work, ask yourself, "How can I use repeated addition to help me find the product?"

5. Lisa put 4 dishes on the table. She put 4 carrots on each dish. How many carrots were there in all?

   Ⓐ 8 carrots

   Ⓑ 18 carrots

   Ⓒ 12 carrots

   Ⓓ 16 carrots

Solve another problem. As you work, ask yourself, "What multiplication sentence do I write for this problem?"

6. Lisa counted 9 pairs of shoes in the hall closet. What is the total number of shoes in the closet?

   Ⓐ 16 shoes

   Ⓑ 18 shoes

   Ⓒ 14 shoes

   Ⓓ 12 shoes

Look at the answer choices for each question.
Read why each answer choice is correct or not correct.

5. Lisa put 4 dishes on the table. She put 4 carrots on each dish. How many carrots were there in all?

Ⓐ 8 carrots

This answer is not correct because $4 + 4 + 4 + 4 = 16$, *not 8. You may have added only* $4 + 4$, *which equals 8.*

Ⓑ 18 carrots

This answer is not correct because $4 + 4 + 4 + 4 = 16$, *not 18. You may not have added correctly.*

Ⓒ 12 carrots

This answer is not correct because $4 + 4 + 4 + 4 = 16$, *not 12. You may have added only* $4 + 4 + 4$, *which equals 12.*

● 16 carrots

This answer is correct because $4 + 4 + 4 + 4 = 16$.

6. Lisa counted 9 pairs of shoes in the hall closet. What is the total number of shoes in the closet?

Ⓐ 16 shoes

This answer is not correct because $9 \times 2 = 18$ *is the correct multiplication sentence for* $2 + 2 + 2 + 2 + 2 + 2 + 2 + 2 + 2 = 18$. *You may have written* $2 + 2 + 2 + 2 + 2 + 2 + 2 + 2 = 16$.

● 18 shoes

This answer is correct because $9 \times 2 = 18$ *is the correct multiplication sentence for* $2 + 2 + 2 + 2 + 2 + 2 + 2 + 2 + 2 = 18$.

Ⓒ 14 shoes

This answer is not correct because $9 \times 2 = 18$ *is the correct multiplication sentence for* $2 + 2 + 2 + 2 + 2 + 2 + 2 + 2 + 2 = 18$. *You may have written* $2 + 2 + 2 + 2 + 2 + 2 + 2 = 14$.

Ⓓ 12 shoes

This answer is not correct because $9 \times 2 = 18$ *is the correct multiplication sentence for* $2 + 2 + 2 + 2 + 2 + 2 + 2 + 2 + 2 = 18$. *You may have written* $2 + 2 + 2 + 2 + 2 + 2 = 12$.

PART FOUR: Learn More About Multiplication

You can skip-count to find a product for a multiplication problem.

- Look at one factor. Count by that number. Skip-count as many times as the other factor.

  For 5×6, skip-count by 5s six times: 5, 10, 15, 20, 25, 30. The last number tells the answer. So, $5 \times 6 = 30$. Since 5×6 has the same product as 6×5, you can skip-count the same way for 6×5.

Skip-count by 5s six times.	**5,**	**10,**	**15,**	**20,**	**25,**	**30**
	(1)	(2)	(3)	(4)	(5)	(6)

- To skip-count, use the smaller factor, or use the factor that is easier to skip-count, like 2, 5, or 10.

Lisa takes the bus to school. She wrote some multiplication problems about her bus ride today. Do Numbers 7 through 10.

7. I met 3 friends on the bus. Each friend has 3 pets. How many pets do they have in all?

Ⓐ 9 pets

Ⓑ 6 pets

Ⓒ 12 pets

Ⓓ 3 pets

8. The bus stopped 8 times. At each stop, 4 children got on. How many children got on the bus?

Ⓐ 24 children

Ⓑ 32 children

Ⓒ 28 children

Ⓓ 36 children

9. There are 7 girls with red gloves. Each girl has 2 red gloves. How many red gloves do the girls have in all?

Ⓐ 12 red gloves

Ⓑ 14 red gloves

Ⓒ 18 red gloves

Ⓓ 16 red gloves

10. The bus has 10 rows of seats. Each row has 4 seats. How many seats are there on the bus?

Ⓐ 14 seats

Ⓑ 20 seats

Ⓒ 44 seats

Ⓓ 40 seats

Read what Lisa wrote about a party. Then do Numbers 11 through 14.

Last summer, we had a party on the Fourth of July. Lots of friends came to our house. The party was in our backyard. We grilled hamburgers, cheeseburgers, and hotdogs. We made a fruit salad. Other people brought food. Everybody played games. Then we waited for it to get dark. When it was dark, we watched the fireworks.

11. Lisa's mom made 8 cheeseburgers. She put 2 slices of cheese on each cheeseburger. How many slices of cheese did she use?

Ⓐ 20 slices

Ⓑ 18 slices

Ⓒ 14 slices

Ⓓ 16 slices

12. Lisa and the other children played a ball game. If someone catches the ball in the game, the person gets 3 points. Lisa caught the ball 9 times. How many points did she get?

Ⓐ 21 points

Ⓑ 27 points

Ⓒ 18 points

Ⓓ 24 points

13. Lisa's neighbor brought 4 trays of cupcakes. There were 6 cupcakes on each tray. How many cupcakes were there in all?

Ⓐ 18 cupcakes

Ⓑ 24 cupcakes

Ⓒ 30 cupcakes

Ⓓ 28 cupcakes

14. Lisa's dad cut up bananas for the fruit salad. He cut each banana into 10 pieces. He used 5 bananas. How many pieces of banana did he put into the salad?

Ⓐ 60 pieces

Ⓑ 45 pieces

Ⓒ 50 pieces

Ⓓ 40 pieces

PART FIVE: Prepare for a Test

▶ A test question about multiplication may ask for the product of two factors.

In art class, Lisa learned about shapes. Read what Lisa learned. Then do Numbers 15 and 16.

All About Shapes

A circle is a ring. It has no straight sides and no corners. A triangle has three straight sides and three corners. A square has four straight sides, and all the sides are the same length. A square also has four corners. A star can have many straight sides and many corners. The corners on a star are called points.

You can use shapes to draw things. A square with a triangle on top is a house. Squares can be windows. A circle can be the doorknob.

Applying Multiplication

15. Lisa looked for shapes in the art room. She found 6 triangles.

How many corners are there in 6 triangles?

Ⓐ 18 corners

Ⓑ 12 corners

Ⓒ 15 corners

Ⓓ 16 corners

Applying Multiplication

16. Lisa drew 7 stars. Each star had 5 points.

How many points were there in all?

Ⓐ 30 points

Ⓑ 35 points

Ⓒ 28 points

Ⓓ 40 points

Read this report that Lisa wrote about teeth. Then do Numbers 17 and 18.

Teeth

Without teeth, it would be very hard to eat. There are different kinds of teeth. At the front of the mouth, there are teeth for biting into food. At the back of the mouth, there are teeth for chewing food.

Newborns have no teeth. Later, babies grow teeth. Some of their teeth will fall out. Then new teeth grow in their place. By the time they are grown up, most people have 32 teeth. There are 16 teeth on the top, and 16 teeth on the bottom.

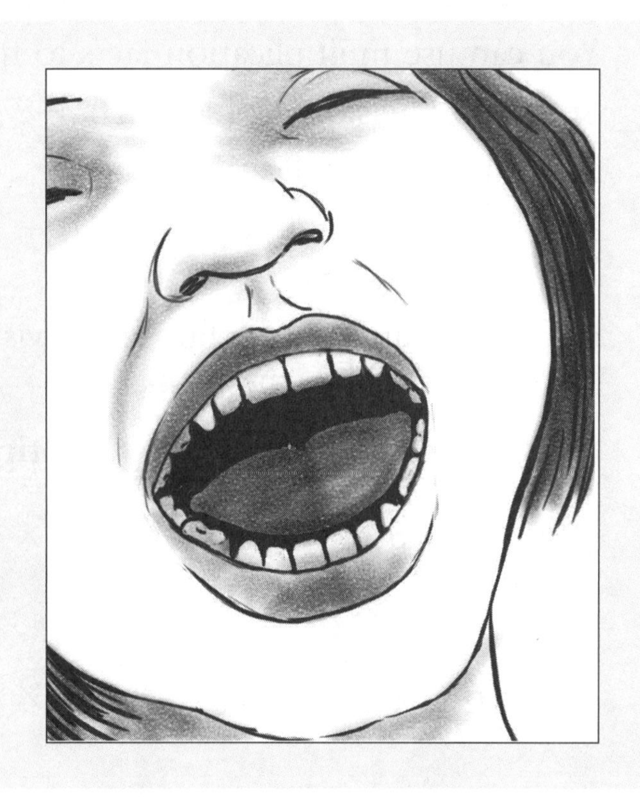

Applying Multiplication

17. Lisa asked 3 friends how many baby teeth they had lost so far. She found out that each friend had lost 7 teeth.

How many teeth had Lisa's friends lost in all?

Ⓐ 18 teeth

Ⓑ 24 teeth

Ⓒ 21 teeth

Ⓓ 26 teeth

Applying Multiplication

18. Lisa counted her own teeth. She has 12 teeth on the top. She has the same number of teeth on the bottom.

How many teeth does Lisa have?

Ⓐ 18 teeth

Ⓑ 24 teeth

Ⓒ 22 teeth

Ⓓ 28 teeth

Strategy Six APPLYING DIVISION

PART ONE: Think About Division

WHAT DO YOU KNOW ABOUT FINDING A QUOTIENT?

You can use multiplication facts to help find a quotient.

Study the division problem: $16 \div 2 = ?$

First, use a multiplication fact: $2 \times$ ___ $= 16 \rightarrow 2 \times \underline{8} = 16$.

Next, use a division sentence: $16 \div 2 = \underline{\ 8\ }$.
The quotient is 8.

▶ Use a multiplication fact and division sentence to help find each quotient. Fill in the blanks.

Problem	Multiplication Fact	Division Sentence
a. $15 \div 5 = ?$	$5 \times$ ____ $= 15$	$15 \div 5 =$ ____
b. $16 \div 4 = ?$	$4 \times$ ____ $= 16$	$16 \div 4 =$ ____
c. $18 \div 6 = ?$	$6 \times$ ____ $= 18$	$18 \div 6 =$ ____
d. $20 \div 4 = ?$	$4 \times$ ____ $= 20$	$20 \div 4 =$ ____
e. $24 \div 2 = ?$	$2 \times$ ____ $= 24$	$24 \div 2 =$ ____
f. $25 \div 5 = ?$	$5 \times$ ____ $= 25$	$25 \div 5 =$ ____

You just reviewed how to use multiplication facts to help find a quotient.

WHAT DO YOU KNOW ABOUT LONG DIVISION?

You can use long division to divide a two-digit dividend by a one-digit divisor.

- Solve each problem using long division. The first problem has been completed for you.

Division Sentence	Long Division
a. $64 \div 4 =$ 16	a. $4\overline{)64}$ = 16; 4; 24; 24; 0
b. $39 \div 3 =$ _____	b.
c. $48 \div 4 =$ _____	c.
d. $72 \div 6 =$ _____	d.
e. $84 \div 4 =$ _____	e.
f. $96 \div 6 =$ _____	f.

You just reviewed how to find a quotient using long division.

Work with a partner.

Together, solve two division problems: $105 \div 5$ and $144 \div 8$. Discuss the way you found each quotient.

PART TWO: Learn About Division

Study the problem that Julio's teacher wrote on the chalkboard. As you study, think about the two ways that Julio solved the problem.

Problem: Dara wants to put the same number of buttons into each cup. She has 15 buttons. She has 3 cups. How many buttons will Dara put into each cup?

Two Ways That Julio Solves the Problem	
1. I draw a picture and count how many in each group.	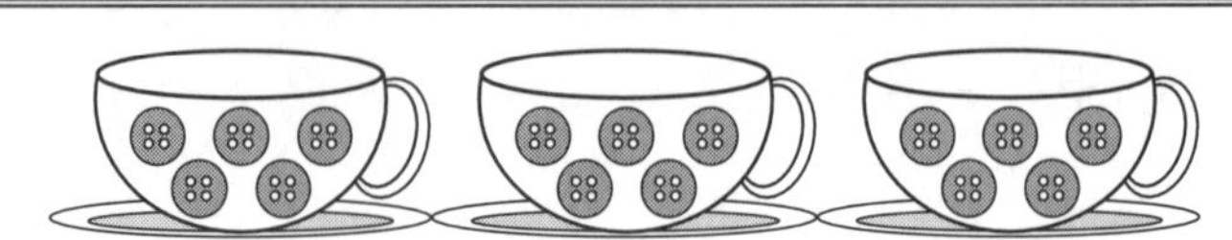
2. I use fact families. I write a multiplication sentence. I rewrite it as a division sentence.	 $3 \times \underline{?} = 15 \rightarrow 3 \times \underline{5} = 15$ $15 \div 3 = 5$

Answer: Dara will put 5 buttons into each cup.

When you divide, you separate a number into equal amounts.

A division problem has three parts. Divide the dividend by the divisor. The answer is the quotient.

$$24 \div 6 = 4$$

dividend divisor quotient

You can write a division problem two ways.

Division Sentence

$28 \div 4 = 7$

Long Division

$$4\overline{)28}$$ with quotient 7

7 — quotient
divisor — 4)28 — dividend

You use **division** to find a quotient.

- To divide, separate a number into equal amounts.
- Use multiplication facts to help find a quotient.

Study another one of Julio's problems. Look at the ways that Julio solved the problem. Then do Numbers 1 through 4.

Problem: Rosa practiced playing her flute for the recital. Each week, she practiced 3 times. She practiced 18 times in all. For how many weeks did Rosa practice?

Separate 18 into 3 equal groups. Count how many in each group.	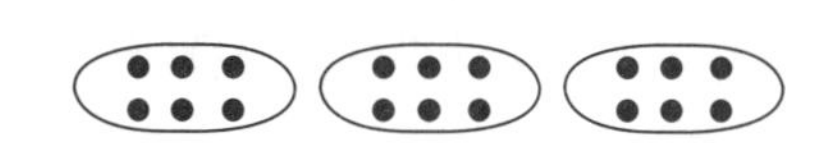
Use multiplication facts to write a multiplication sentence. Rewrite it as a division sentence.	$3 \times \underline{?} = 18 \rightarrow 3 \times \underline{6} = 18$ $18 \div 3 = 6$

Answer: Rosa practiced for 6 weeks.

1. The recital lasted for 36 minutes. Each child, including Julio, played for 4 minutes. How many children were in the recital?

   Ⓐ 7 children
   Ⓑ 9 children
   Ⓒ 6 children
   Ⓓ 8 children

2. Julio tried out drum sets at the music store. There were 3 drums in each set. In all, he tried 15 drums. How many sets did Julio try?

   Ⓐ 3 sets
   Ⓑ 5 sets
   Ⓒ 8 sets
   Ⓓ 6 sets

3. Julio played the drums for 4 minutes. Each song he played was 2 minutes long. How many songs did Julio play?

   Ⓐ 2 songs
   Ⓑ 3 songs
   Ⓒ 4 songs
   Ⓓ 8 songs

4. Julio plays the piano for 10 hours each week. He plays for 2 hours at a time. How many times does Julio play the piano each week?

   Ⓐ 6 times
   Ⓑ 4 times
   Ⓒ 3 times
   Ⓓ 5 times

Talk about your answers to questions 1–4. Tell why you chose the answers you did.

PART THREE: Check Your Understanding

Remember: You use division to find a quotient.

- To divide, separate a number into equal amounts.
- Use multiplication facts to help find a quotient.

Solve this problem. As you work, ask yourself, "How do I separate this number into equal amounts?"

5. Julio invited 14 friends to a party. Julio and his sister wrote the invitations. If they each wrote the same number of invitations, how many did each child write?

Ⓐ 9 invitations

Ⓑ 7 invitations

Ⓒ 6 invitations

Ⓓ 8 invitations

Solve another problem. As you work, ask yourself, "How can I use multiplication facts to help me find the quotient?"

6. At the party, Julio, his family, and his guests played tag. They formed 4 equal teams. In all, 20 people played the game. How many people were on each team?

Ⓐ 4 people

Ⓑ 6 people

Ⓒ 5 people

Ⓓ 3 people

Look at the answer choices for each question.
Read why each answer choice is correct or not correct.

5. Julio invited 14 friends to a party. Julio and his sister wrote the invitations. If they each wrote the same number of invitations, how many did each child write?

Ⓐ 9 invitations

This answer is not correct because 14 divided into 2 equal amounts is 7, not 9.

● 7 invitations

This answer is correct because 14 divided into 2 equal amounts is 7.

Ⓒ 6 invitations

This answer is not correct because 14 divided into 2 equal amounts is 7, not 6.

Ⓓ 8 invitations

This answer is not correct because 14 divided into 2 equal amounts is 7, not 8.

6. At the party, Julio, his family, and his guests played tag. They formed 4 equal teams. In all, 20 people played the game. How many people were on each team?

Ⓐ 4 people

This answer is not correct because $4 \times 5 = 20$, which you can rewrite as $20 \div 4 = 5$. Multiplication facts tell you that $4 \times 4 = 16$, not 20.

Ⓑ 6 people

This answer is not correct because $4 \times 5 = 20$, which you can rewrite as $20 \div 4 = 5$. Multiplication facts tell you that $4 \times 6 = 24$, not 20.

● 5 people

This answer is correct because $4 \times 5 = 20$, which you can rewrite as $20 \div 4 = 5$.

Ⓓ 3 people

This answer is not correct because $4 \times 5 = 20$, which you can rewrite as $20 \div 4 = 5$. Multiplication facts tell you that $4 \times 3 = 12$, not 20.

PART FOUR: Learn More About Division

You can use long division to divide a two-digit dividend by a one-digit divisor.

▶ Divide the tens. Then divide the ones. Check your answer by multiplying the quotient by the divisor.

Line up the tens and ones in the quotient with the tens and ones in the dividend.

4)84 1. Divide the tens: 4 goes into 8 2 times. Multiply. 4 × 2 = 8 Subtract. 8 − 8 = 0 2 4)84 − 8 0	2. Bring down the ones digit. 2 4)84 − 8 04	3. Divide the ones: 4 goes into 4 1 time. Multiply. 4 × 1 = 4 Subtract. 4 – 4 = 0 21 4)84 − 8 04 − 4 0	4. Check your answer. 21 × 4 84

Julio went to the mall with his family. Do Numbers 7 through 10.

7. Julio and his brothers counted cars in the parking lot. They counted 88 cars. The cars were parked in 4 rows. Each row had the same number of cars. How many cars were in each row?

Ⓐ 11 cars Ⓒ 18 cars

Ⓑ 43 cars Ⓓ 22 cars

8. Julio's family ate lunch in the mall. The restaurant had 48 chairs. There were 4 chairs at each table. How many tables were there?

Ⓐ 10 tables Ⓒ 14 tables

Ⓑ 11 tables Ⓓ 12 tables

9. Julio likes to look at the puppies in the pet store. He counted 20 puppies in all. There were 2 puppies in each cage. How many cages were there?

Ⓐ 10 cages Ⓒ 9 cages

Ⓑ 8 cages Ⓓ 5 cages

10. After lunch, Julio and his brother spent 60 minutes shopping. The boys spent 5 minutes in each store. How many stores did they visit?

Ⓐ 15 stores Ⓒ 12 stores

Ⓑ 10 stores Ⓓ 30 stores

Julio's class played a counting game for math. Read the rules of the game. Then do Numbers 11 through 14.

Counting Game Rules

A. Teams of students will count items in the school.

B. There will be 3 students on each team.

C. Ms. Heinz will give each team a list that tells what to count.

D. Each team will have 15 minutes to count the items on the list.

E. Teams should write their answers on their list.

F. The first team to turn in a list with all the items counted correctly will win a prize.

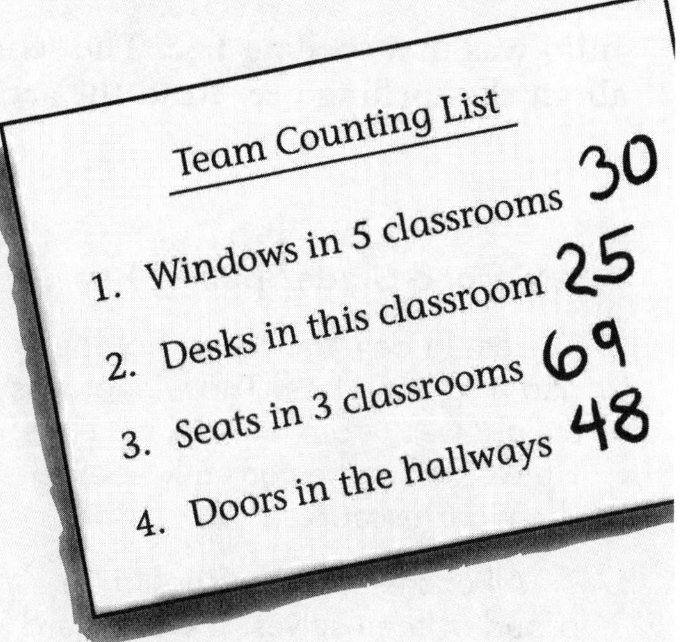

Team Counting List
1. Windows in 5 classrooms 30
2. Desks in this classroom 25
3. Seats in 3 classrooms 69
4. Doors in the hallways 48

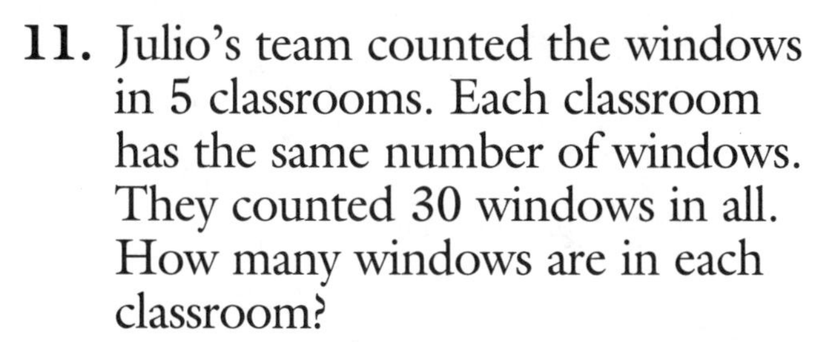

11. Julio's team counted the windows in 5 classrooms. Each classroom has the same number of windows. They counted 30 windows in all. How many windows are in each classroom?

Ⓐ 8 windows　Ⓒ 9 windows

Ⓑ 5 windows　Ⓓ 6 windows

12. Julio and Shantal counted 25 desks in their classroom. Each row has 5 desks. How many rows of desks are there?

Ⓐ 5 rows　Ⓒ 4 rows

Ⓑ 9 rows　Ⓓ 6 rows

13. Julio's team counted the seats in 3 classrooms. Each classroom has the same number of seats. There are 69 seats in all. How many seats are in each classroom?

Ⓐ 31 seats　Ⓒ 23 seats

Ⓑ 22 seats　Ⓓ 13 seats

14. Julio's friend Ming counted 48 doors in the hallways. The same number of doors are in each of the 4 hallways. How many doors are in each hallway?

Ⓐ 6 doors　Ⓒ 10 doors

Ⓑ 12 doors　Ⓓ 8 doors

- A test question about division may ask for a quotient.
- A test question about division may ask for the quotient of a two-digit dividend divided by a one-digit divisor.

Julio was in a spelling bee. The school newsletter had a short article about the spelling bee. Read the article. Then do Numbers 15 and 16.

Second-Grade Spelling Bee

Last Friday, the second graders had a spelling bee. The winner was Julio Paseo, from Ms. Heinz's class. He won when he correctly spelled the word *escape*.

All of the students should be proud of themselves. It was a hard contest, and every student tried his or her best.

Special thanks to the parents who helped out with this event.

Applying Division

15. Before the spelling bee, Julio and the other children set up chairs for the audience. Julio set up 27 chairs. He lined the chairs up in 3 equal rows.

How many chairs did Julio put in each row?

Ⓐ 6 chairs　Ⓒ 9 chairs
Ⓑ 8 chairs　Ⓓ 5 chairs

Applying Division

16. There were 5 students in the final round of the spelling bee. Each student had to spell the same number of words. In all, 20 words were spelled in the final round.

How many words did each student spell in the final round?

Ⓐ 4 words　Ⓒ 5 words
Ⓑ 7 words　Ⓓ 8 words

Read Julio's journal entry about his visit to a farm. Then do Numbers 17 and 18.

Tuesday

At the farm, I saw cows, chickens, horses, and goats. I also saw a tractor, a bull, and a rooster. I helped to milk a cow, and I even got to taste fresh milk! The rooster was very loud when it crowed, but I did not think it sounded like "cock-a-doodle-doo." To me, it sounded like "ra-ra-roooo!" My favorite part of the day was feeding carrots to the horses. I had fun at the farm. Some day, maybe I will be a farmer.

Applying Division

17. Julio counted 55 cows at the farm. The cows were in 5 different fields. The same number of cows were in each field.

How many cows were in each field?

Ⓐ 11 cows

Ⓑ 13 cows

Ⓒ 17 cows

Ⓓ 15 cows

Applying Division

18. The farmer told Julio that she collected 3 eggs from each hen last week. She said that she collected a total of 93 eggs that week. Julio figured out how many hens the farmer had.

What was Julio's correct answer?

Ⓐ 26 hens

Ⓑ 31 hens

Ⓒ 33 hens

Ⓓ 13 hens

Strategies Four–Six REVIEW

PART ONE: Read a Story

Grandpa told Shelly and her brother a story. Read what happened next. Then do Numbers 1 through 6.

A Sweet Idea

"Once, there was a special cottage deep in the forest. The cottage walls were made of gingerbread. The windows were made of frosting. The roof was made of gumdrops, and the walkway was made of jellybeans. The doorknobs were chocolate chips!

One day, two children found the cottage . . ."

As Grandpa read, Shelly had an idea. She was going to make her own cottage! When Grandpa finished telling the story, Shelly asked Grandma to help her.

First, Shelly and her grandmother went to the store. They bought gumdrops, jellybeans, chocolate chips, and peppermint sticks. They also bought frosting and graham crackers.

When they got home, everyone went to the kitchen. Shelly took the graham crackers out of the box. She picked 4 large squares for the walls of the house. She broke another square into 2 rectangles for the roof. Shelly gave the gumdrops to her brother. She gave the jellybeans to Grandma. Grandpa got the frosting. Shelly took the chocolate chips.

Shelly and her brother held the wall and roof pieces in place while Grandpa used frosting to stick them together. They used more frosting to stick gumdrops and jellybeans to the roof and walls of the house. Then Grandpa dipped a toothpick into the frosting and drew windows and a door. Grandma used peppermint sticks to make a fence around the house. Finally, Shelly put on a chocolate-chip doorknob.

"Now let's eat it!" said Shelly.

"Oh, no!" said her brother. "The two children who found the gingerbread cottage tried to eat the whole thing. They got terrible stomach pains!"

"Oh, no," said Shelly. "Let's put *this* cottage on the shelf!"

Applying Subtraction

1. There were 12 graham crackers in the box. Shelly used 5 crackers for the walls and roof of the cottage.

   How many crackers were left?

   Ⓐ 7 crackers Ⓒ 5 crackers
   Ⓑ 6 crackers Ⓓ 8 crackers

Applying Subtraction

2. Shelly and her brother put a total of 65 gumdrops and jellybeans on the cottage. In all, they used 37 gumdrops.

   How many jellybeans did they use?

   Ⓐ 32 jellybeans Ⓒ 25 jellybeans
   Ⓑ 28 jellybeans Ⓓ 38 jellybeans

Applying Multiplication

3. Grandpa made 2 frosting windows on each wall. There were 4 walls.

   How many windows did Grandpa make in all?

   Ⓐ 6 windows
   Ⓑ 9 windows
   Ⓒ 8 windows
   Ⓓ 7 windows

Applying Multiplication

4. The 4 graham-cracker walls were each 6 inches long.

   What would be the total length of the 4 crackers?

   Ⓐ 18 inches Ⓒ 16 inches
   Ⓑ 24 inches Ⓓ 20 inches

Applying Division

5. The roof had a total of 18 gumdrops in 3 equal rows.

   How many gumdrops were in each row?

   Ⓐ 9 gumdrops
   Ⓑ 5 gumdrops
   Ⓒ 8 gumdrops
   Ⓓ 6 gumdrops

Applying Division

6. Grandma used 4 peppermint sticks to make the fence. The total length of the fence was 20 inches.

   How long was each peppermint stick?

   Ⓐ 5 inches Ⓒ 8 inches
   Ⓑ 6 inches Ⓓ 4 inches

PART TWO: Read an Article

Read the article about inventions. Then do Numbers 7 through 12.

People have always invented things. Some inventions, like the wheel, have changed history. Other inventions have not been quite so successful. For example, in 1903, an American farmer invented eyeglasses for chickens. He hoped the glasses would protect the chickens from pecking one another in the eyes. But chickens just don't like to wear glasses!

Usually, an inventor wants to make something that will help people. Alexander Graham Bell invented the telephone for people who lived far away from each other. With a telephone, they could talk to each other whenever they wanted. They did not have to write and send a letter.

Sometimes, an invention happens by accident. That happened with Richard James. In 1943, he was trying to invent a tool for ships to use. A spring fell off his desk and seemed to "walk" down a pile of books. He took the spring home and showed the trick to his children. They loved it! And millions of children since then have also loved the Slinky®!

The invention of chewing gum was also an accident. In the 1870s, Thomas Adams was trying to make rubber. He was using something called *chicle*. Chicle is the dried sap of a tree. Adams's son kept chewing on pieces of chicle from his father's lab. When Adams saw how much his son enjoyed the treat, he gave up his plan to make rubber. Instead, he made chewing gum. Today, chewing gum comes in many shapes, sizes, and flavors.

Applying Subtraction

7. There are 12 buttons on Ari's telephone. There are numbers on 10 of the buttons.

   How many buttons without numbers are on Ari's telephone?

   Ⓐ 6 buttons
   Ⓑ 4 buttons
   Ⓒ 3 buttons
   Ⓓ 2 buttons

Applying Subtraction

8. Maria blew a bubble that was 25 centimeters across. Jack blew a bubble that was 18 centimeters across.

   How much bigger was Maria's bubble than Jack's?

   Ⓐ 7 centimeters
   Ⓑ 5 centimeters
   Ⓒ 13 centimeters
   Ⓓ 17 centimeters

Applying Multiplication

9. Each pair of eyeglasses has 2 lenses.

   How many lenses would be in 16 pairs of eyeglasses?

   Ⓐ 28 lenses
   Ⓑ 32 lenses
   Ⓒ 36 lenses
   Ⓓ 26 lenses

Applying Multiplication

10. Ji Lee bought 8 packs of chewing gum. Each pack has 5 pieces of gum.

    How many pieces of gum did Ji Lee buy in all?

    Ⓐ 35 pieces
    Ⓑ 30 pieces
    Ⓒ 40 pieces
    Ⓓ 45 pieces

Applying Division

11. A metal Slinky® is made of about 80 inches of coiled wire.

    If the wire were cut into 5 equal pieces, how long would each piece be?

    Ⓐ 12 inches
    Ⓑ 16 inches
    Ⓒ 14 inches
    Ⓓ 10 inches

Applying Division

12. Alan visited the car dealer with his mother. He counted a total of 84 wheels on the cars.

    If each car has 4 wheels, how many cars were at the car dealer's?

    Ⓐ 16 cars   Ⓒ 19 cars
    Ⓑ 20 cars   Ⓓ 21 cars

Strategy Seven CONVERTING TIME AND MONEY

PART ONE: Think About Time and Money

WHAT DO YOU KNOW ABOUT TIME?

A clock has one face and two hands.

The face has the numbers 1 through 12.

The shorter hand on the clock is the hour hand. It takes 60 minutes for the hour hand to move from one number to the next number.

The longer hand on a clock is the minute hand. It takes 5 minutes for the minute hand to move from one number to the next number.

▶ Place the two hands on each clock face to show the time.

a. The time is 4 o'clock.

b. The time is 6:30 o'clock.

c. The time is 12:15 o'clock.

d. The time is 5:45 o'clock.

e. The time is 8:00 o'clock.

f. The time is 3:20 o'clock.

You just reviewed how to show hours and minutes on a clock face.

WHAT DO YOU KNOW ABOUT COINS AND THEIR VALUE?

The value of a quarter is 25¢. 

The value of a dime is 10¢. 

The value of a nickel is 5¢. 

The value of a penny is 1¢. 

▶ Study each group of coins. Write the value of each group.

a.   

The amount of money is ________¢.

b.   

The amount of money is ________¢.

c.  

The amount of money is ________¢.

d.  

The amount of money is ________¢.

You just reviewed information about coins and their value.

Together, figure out ten ways that you can show coins that have a value of 76¢. Write and discuss the ten ways that show the amount.

Study the clock. Read what Sandy learned about clocks.
As you study, think about how to use a clock to tell time.

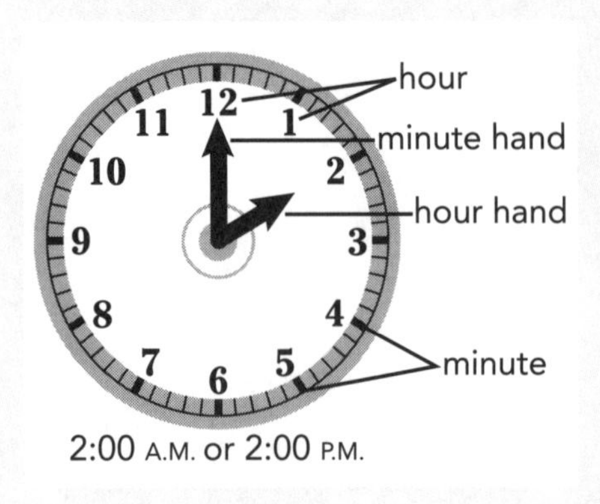

2:00 A.M. or 2:00 P.M.

A clock has two hands.

- The shorter hand is the hour hand.
  The hour hand takes 60 minutes to move from one number to the next.
- The longer hand is the minute hand.
  The minute hand tells how many minutes after or before an hour. The minute hand takes 5 minutes to move from one number to the next number.

Look at *Clock A* below. It shows 4:15. Count by 1s from 12 to find that four hours have gone by. Skip-count by 5s to find that 15 minutes have gone by.

Look at *Clock B* below. It shows 6:30. To find how many hours have gone by from *Clock A* to *Clock B*, count by 1s from 4 to 6 and get 2. Then skip-count by 5s from 15 to 30 and get 15. So, 2 hours, 15 minutes have passed.

Clock B
6:30

You use a clock to tell **time** and to tell how much time has gone by.

- Look at the shorter hand on the clock to tell the hour. Look at the longer hand to tell how many minutes there are after or before an hour.
- Count by 1s to find how many hours have gone by. Skip-count by 5s to find how many minutes have gone by.
- To write time, use A.M. to show morning time. Use P.M. to show afternoon and evening time.

Study how Sandy solved this problem. Then do Numbers 1 through 4.

Problem: The first clock shows the time a boy wakes up in the morning. The boy gets to school 1 hour and 40 minutes after he wakes up. The second clock shows the time he gets to school. What time does he wake up? What time does he get to school?

1. I see the hour hand is between 7 and 8. I count by 5s from 12 to find how many minutes after 7:00 the boy wakes up.	2. I see the hour hand is on 9. I know that when the minute hand is on 12, the time is exactly on the hour.
 Wakes Up	 Gets to School

Answer: He wakes up at 7:20 A.M.

Answer: He gets to school at 9:00 A.M.

1. Sandy's school starts at the time shown. When does school start?

Ⓐ 8:00 A.M.
Ⓑ 5:30 A.M.
Ⓒ 7:00 A.M.
Ⓓ 12:40 A.M.

2. Sandy's school ends at the time shown. When does school end?

Ⓐ 5:00 P.M.
Ⓑ 7:30 P.M.
Ⓒ 3:00 P.M.
Ⓓ 12:15 P.M.

3. Look at the clocks in Numbers 1 and 2. How long is Sandy in school?

Ⓐ 5 hours
Ⓑ 7 hours
Ⓒ 3 hours and 15 minutes
Ⓓ 10 hours

4. Look at the clock in Number 2. Sandy gets home 15 minutes later. What time does Sandy get home?

Ⓐ 5:45 P.M.   Ⓒ 4:00 P.M.
Ⓑ 12:45 P.M.   Ⓓ 3:15 P.M.

Talk about your answers to questions 1–4.
Tell why you chose the answers you did.

PART THREE: Check Your Understanding

Remember: You use a clock to tell time and to tell how much time has gone by.

- Look at the shorter hand on the clock to tell the hour. Look at the longer hand to tell how many minutes there are after or before an hour.
- Count by 1s to find how many hours have gone by. Skip-count by 5s to find how many minutes have gone by.
- To write time, use A.M. to show morning time. Use P.M. to show afternoon and evening time.

Solve this problem. As you work, ask yourself, "Where does the hour hand point? Where does the minute hand point?"

5. On Saturday morning, Sandy had her teeth cleaned at the time shown on the clock. What time did Sandy have her teeth cleaned?

Ⓐ 2:00 P.M.
Ⓑ 11:30 A.M.
Ⓒ 1:15 P.M.
Ⓓ 9:00 A.M.

Solve another problem. As you work, ask yourself, "How can counting by 1s and skip-counting by 5s help me know how much time has gone by?"

6. Later, Sandy and her mom went to dinner and a movie. The clocks show when they left home and when they came back. How long were they gone?

Ⓐ 3 hours and 20 minutes
Ⓑ 1 hour
Ⓒ 2 hours
Ⓓ 5 hours and 30 minutes

Look at the answer choices for each question.
Read why each answer choice is correct or not correct.

5. On Saturday morning, Sandy had her teeth cleaned at the time shown on the clock. What time did Sandy have her teeth cleaned?

Ⓐ 2:00 P.M.

This answer is not correct because when it is 2:00, the hour hand points to 2 and the minute hand points to 12.

● 11:30 A.M.

This answer is correct because the hour hand is between 11 and 12, and the minute hand points to 6, which shows that it is 30 minutes after 11, or 11:30. The A.M. shows morning time.

Ⓒ 1:15 P.M.

This answer is not correct because when it is 1:15, the hour hand is between 1 and 2, and the minute hand points to 3, which shows that it is 15 minutes after the hour.

Ⓓ 9:00 A.M.

This answer is not correct because when it is 9:00, the hour hand points to 9 and the minute hand points to 12.

6. Later, Sandy and her mom went to dinner and a movie. The clocks show when they left home and when they came back. How long were they gone?

● 3 hours and 20 minutes

This answer is correct because if you count hours by 1s and minutes by 5s, you get 3 hours and 20 minutes.

Ⓑ 1 hour

This answer is not correct because it is less than the right amount of time.

Ⓒ 2 hours

This answer is not correct because it is less than the right amount of time.

Ⓓ 5 hours and 30 minutes

This answer is not correct because it is more than the right amount of time. If you count hours by 1s and minutes by 5s, you get 3 hours and 20 minutes.

PART FOUR: Learn About Money

You use what you know about the value of coins to count **money**.

▶ To find the total value of a group of coins, count all the coins together. Start with the coins with the highest value.

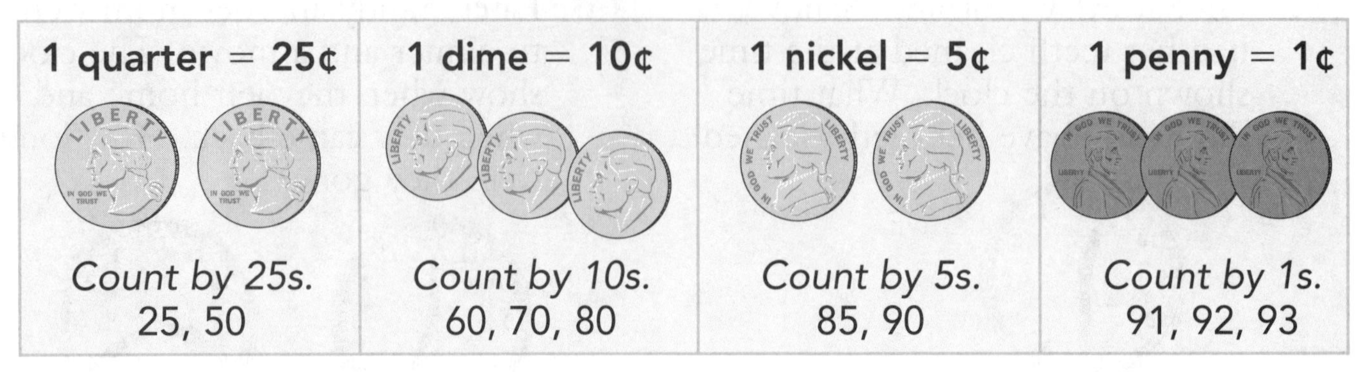

1 quarter = 25¢	1 dime = 10¢	1 nickel = 5¢	1 penny = 1¢
Count by 25s. 25, 50	*Count by 10s.* 60, 70, 80	*Count by 5s.* 85, 90	*Count by 1s.* 91, 92, 93

The total value of the coins shown in the chart is 93¢.

Sandy and her mom went to a flea market. Do Numbers 7 through 10.

7. Sandy bought a doll for the coins shown. How much did she pay?

Ⓐ 32¢ Ⓒ 27¢

Ⓑ 22¢ Ⓓ 37¢

8. Sandy paid 60¢ for a used book. What group totals 60¢?

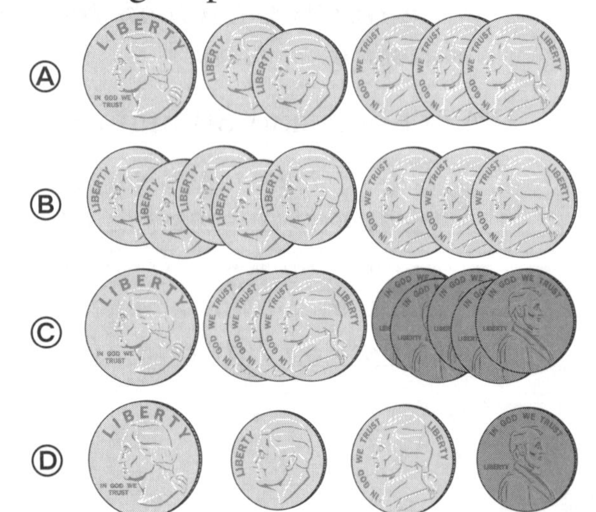

9. Sandy bought a dish for the coins shown. How much did she pay?

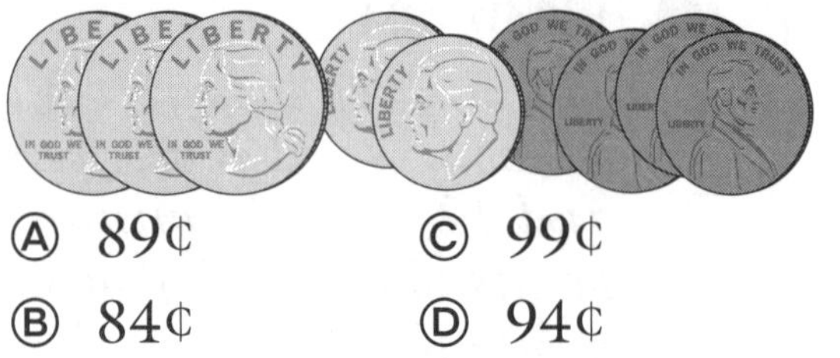

Ⓐ 89¢ Ⓒ 99¢

Ⓑ 84¢ Ⓓ 94¢

10. Sandy paid 55¢ for a stuffed tiger. What group does *not* total 55¢?

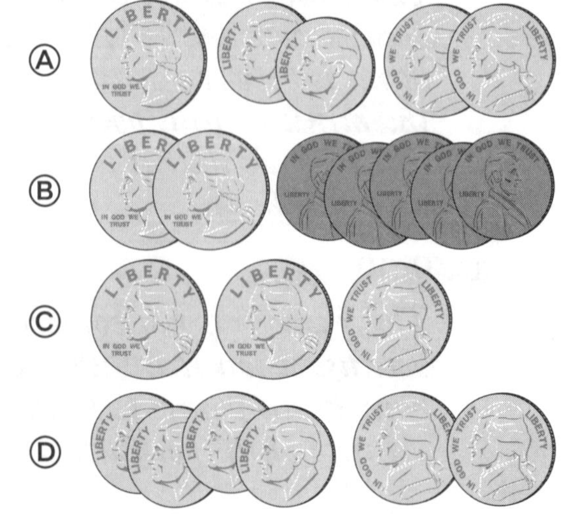

Read what Sandy learned about the past from her grandmother. Then do Numbers 11 through 14.

On Sandy's birthday, her grandmother gave her a Slinky®. "It was once my favorite toy," she said. "Richard James invented the Slinky. The Slinky was first sold in 1945, the year I was born. Then, the Slinky cost \$1.99. Today, you can still get a Slinky for about that price."

11. Grandmother paid \$2.39 for the new Slinky. That is 40¢ more than the first Slinky cost. What group of coins does *not* total 40¢?

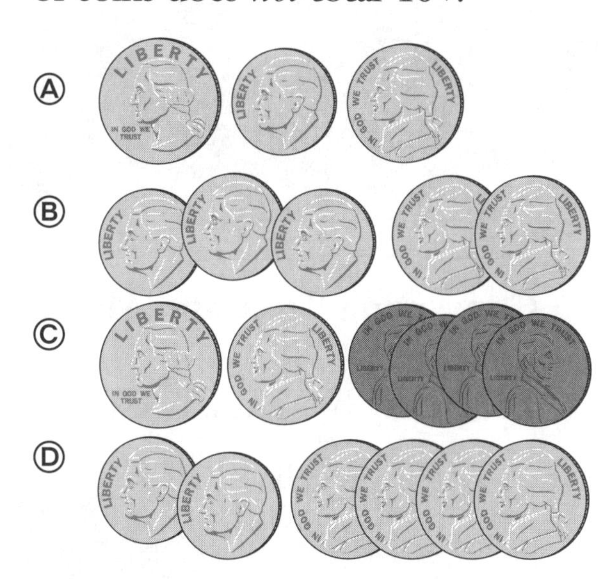

Ⓐ

Ⓑ

Ⓒ

Ⓓ

12. When Grandmother was a child, bread cost the same as the coins shown. How much did bread cost?

Ⓐ 30¢

Ⓑ 35¢

Ⓒ 25¢

Ⓓ 20¢

13. Sandy's grandmother took a bus to the birthday party. The bus fare was 85¢. What group totals 85¢?

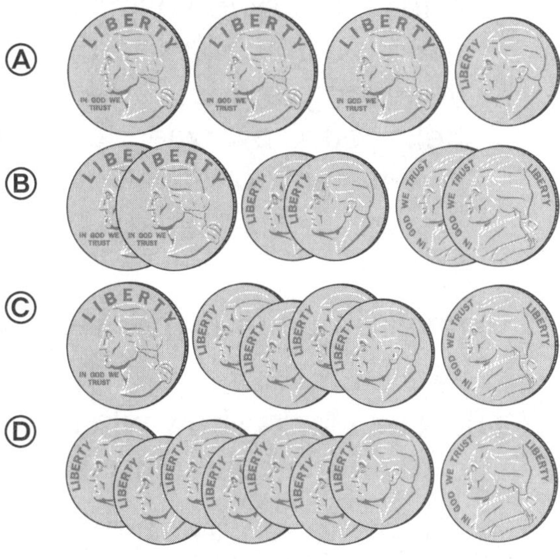

Ⓐ

Ⓑ

Ⓒ

Ⓓ

14. Sandy's grandmother said that she used to go to a movie for the coins shown. How much did it cost to see a movie?

Ⓐ 13¢

Ⓑ 28¢

Ⓒ 23¢

Ⓓ 18¢

PART FIVE: Prepare for a Test

- A test question about time may ask for the time shown on a clock.
- A test question about time may ask for how many hours or minutes have gone by.
- A test question about money may ask for the total value of coins.

Sandy went to the Children's Museum. Read about Sandy's favorite exhibit. Then do Numbers 15 and 16.

A Day of Fun and Learning

A guide led Sandy through a giant model of a mouth. On the make-believe tongue, Sandy felt lots of bumps. "Those are taste buds," the guide explained. "You have bumps like that on your tongue, too. Taste buds help you know when foods are sweet, salty, or sour."

Converting Time and Money

15. The clocks show when Sandy got to and left the museum.

How long was Sandy at the museum?

Ⓐ 5 hours
Ⓑ 3 hours
Ⓒ 4 hours
Ⓓ 6 hours

Converting Time and Money

16. Sandy bought a book for 69¢.

What group totals 69¢?

Ⓐ
Ⓑ
Ⓒ
Ⓓ

Read about Sandy's cousin Corey. Then do Numbers 17 and 18.

Letter from Alaska

Sandy got a large envelope from Alaska. Inside, there was a letter from her cousin Corey and a picture of a children's choir. Corey has been a member of the Northern Children's Choir since she was 5. Her first year, she was in the Young Singers group. Then, when she turned 7, Corey moved up to the Grand Singers group. "We travel all over Alaska," Corey wrote in her letter. "Maybe someday, we'll travel south and sing for you."

Converting Time and Money

17. In her letter, Corey said that she practices with the choir every Saturday morning. Practice starts at the time shown on the clock. It ends 1 hour and 30 minutes later.

What time does practice end?

Ⓐ 11:25 A.M. Ⓒ 10:45 A.M.

Ⓑ 4:00 P.M. Ⓓ 12:05 P.M.

Converting Time and Money

18. The value of the stamps on the envelope Corey sent Sandy is the same as the total value of the coins shown.

How much did Corey spend for the stamps?

Ⓐ 77¢ Ⓒ 67¢

Ⓑ 62¢ Ⓓ 72¢

Strategy Eight

CONVERTING CUSTOMARY AND METRIC MEASURES

PART ONE: Think About Customary and Metric Measures

WHAT DO YOU KNOW ABOUT UNITS OF MEASUREMENT?

To measure liquids, use cups, pints, quarts, or liters.

To measure length, height, and width, use inches, feet, centimeters, and meters.

▶ Convert each measure to one of equal value.

a. 12 inches = ____ foot

b. 36 inches = ____ feet

c. 3 feet = ____ yard

d. 2 cups = ____ pint

e. 2 pints = ____ quart

f. 4 quarts = ____ gallon

You just reviewed equal measures.

To add or subtract measurements that are given in different units, first change them to the same unit.

▶ Complete each addition or subtraction problem. Label each answer.

a. 1 gallon − 2 quarts = ______

b. 3 feet + 20 inches = ______

c. 2 quarts − 2 pints = ______

d. 2 yards + 4 feet = ______

e. 1 gallon + 1 pint = ______

You just reviewed adding and subtracting unlike measures.

WHAT DO YOU KNOW ABOUT ESTIMATING MEASURES?

▶ Think carefully about each measurement sentence. Then write T for TRUE or F for FALSE beside each measurement sentence.

a. Your hand is about 6 meters wide. ______________

b. A page in this book is about 11 inches long. ______________

c. A regular paper clip is nearly 10 centimeters long. ______________

d. A liter is a little more than a quart. ______________

e. The door of your bedroom is about 30 inches wide. ______________

f. A quart is equal to 3 pints. ______________

g. The length of a new pencil is about $7\frac{1}{2}$ inches. ______________

h. There are 200 centimeters in one meter. ______________

i. One meter is about 100 inches. ______________

j. A dollar bill is about 15 centimeters long. ______________

You just reviewed estimating the measure of familiar items.

Together, write four problems about estimating sizes or amounts of familiar items. Then discuss how you figured out each estimate and its unit.

PART TWO: Learn About Customary and Metric Measures

Study the measurement chart that Al's teacher made. As you study, think about the different ways that things can be measured.

UNITS OF MEASUREMENT	
Customary	**Metric**
inch, foot, yard, mile	centimeter, meter
cup, pint, quart, gallon	liter

You use different units of measurement to measure different things.
To measure length, height, or width, use units like inches, feet, or meters.
To measure "how much" of a liquid or a solid, use units like cups, pints, or liters.

Here are some measurement facts.

12 inches = 1 foot
36 inches = 3 feet
3 feet = 1 yard

2 cups = 1 pint
2 pints = 1 quart
4 quarts = 1 gallon

If measurements are given in different units, first use the measurement facts to change them to the same unit. Then you can add or subtract.

For example, 1 foot = 12 inches. To subtract 3 inches from 1 foot, first change 1 foot to 12 inches. Then subtract 3 inches from 12 inches to get 9 inches.

You use **measurement** to find the size of something.

- To measure liquids, use cups, pints, quarts, and liters. To measure length, height, or width, use inches, feet, centimeters, or meters.
- To add or subtract measurements that are given in different units, first change them to the same unit.

Al made a chart of what he measured. Study Al's chart. Think about what units are used to measure each item. Then do Numbers 1 through 4.

WHAT I MEASURED	Breakfast Cereal	Orange Juice	Height of Table	Length of Spoon	My Height
MEASUREMENT	1 cup	2 cups	36 inches	7 inches	3 feet, 10 inches

1. Al had 4 cups of orange juice today. How many pints of orange juice did Al have?

   Ⓐ 3 pints
   Ⓑ 2 pints
   Ⓒ 4 pints
   Ⓓ 1 pint

2. Al measured himself. He found that he was more than 3 feet tall. How many inches are in 3 feet?

   Ⓐ 48 inches
   Ⓑ 16 inches
   Ⓒ 36 inches
   Ⓓ 40 inches

3. Al figured out how tall he would be if he were 1 foot taller. What was his correct answer?

   Ⓐ 58 inches
   Ⓑ 50 inches
   Ⓒ 4 feet
   Ⓓ 53 inches

4. There was 1 quart of orange juice. For breakfast, Al drank 2 cups of the juice. How much orange juice was left?

   Ⓐ 2 cups
   Ⓑ 4 cups
   Ⓒ 1 cup
   Ⓓ 3 cups

Talk about your answers to questions 1–4.
Tell why you chose the answers you did.

PART THREE: Check Your Understanding

Remember: You use measurement to find the size of something.

- To measure liquids, use cups, pints, quarts, and liters. To measure length, height, or width, use inches, feet, or meters.
- To add or subtract measurements that are given in different units, first change them to the same unit.

Solve this problem. As you work, ask yourself, "What measurement facts do I use?"

5. Al has a toy train. He measured its length. He found that the length of the train was 48 inches. How many feet long is Al's train?

   Ⓐ 3 feet
   Ⓑ 24 feet
   Ⓒ 4 feet
   Ⓓ 2 feet

Solve another problem. As you work, ask yourself, "How do I subtract these measurements?"

6. A section of Al's train track was 1 foot long. The width was 10 inches less than the length. What was the width?

   Ⓐ 2 inches
   Ⓑ 4 inches
   Ⓒ 8 inches
   Ⓓ 11 inches

Look at the answer choices for each question.
Read why each answer choice is correct or not correct.

5. Al has a toy train. He measured its length. He found that the length of the train was 48 inches. How many feet long is Al's train?

Ⓐ 3 feet

This answer is not correct because 12 inches = 1 foot. That fact also tells you that 36 inches = 3 feet, so 48 inches cannot also equal 3 feet.

Ⓑ 24 feet

This answer is not correct because 12 inches = 1 foot. That fact also tells you that 24 inches = 2 feet, 36 inches = 3 feet, and 48 inches = 4 feet, not 24 feet.

● 4 feet

This answer is correct because 12 inches = 1 foot. That fact also tells you that 24 inches = 2 feet, 36 inches = 3 feet, and 48 inches = 4 feet.

Ⓓ 2 feet

This answer is not correct because 12 inches = 1 foot. That fact tells you that 24 inches = 2 feet, so 48 inches cannot also equal 2 feet.

6. A section of Al's train track was 1 foot long. The width was 10 inches less than the length. What was the width?

● 2 inches

This answer is correct. To subtract 10 inches from 1 foot, change 1 foot to 12 inches. Then subtract. So, 12 inches − 10 inches = 2 inches.

Ⓑ 4 inches

This answer is not correct. To subtract 10 inches from 1 foot, change 1 foot to 12 inches. Then subtract. So, 12 inches − 10 inches = 2 inches, not 4 inches.

Ⓒ 8 inches

This answer is not correct. To subtract 10 inches from 1 foot, change 1 foot to 12 inches. Then subtract. So, 12 inches − 10 inches = 2 inches, not 8 inches.

Ⓓ 11 inches

This answer is not correct. To subtract 10 inches from 1 foot, change 1 foot to 12 inches. Then subtract. So, 12 inches − 10 inches = 2 inches, not 11 inches.

PART FOUR: Learn More About Customary and Metric Measures

You can estimate the measurement of something without measuring it exactly. Here are some examples that you can use to compare.

- A doorknob is about 1 meter or 1 yard or 36 inches from the bottom of a door.
- A postage stamp is about 1 inch or $2\frac{1}{2}$ centimeters long.
- Your small finger is about 1 centimeter wide.
- The long side of a regular piece of paper is about 1 foot in length.
- A container of yogurt or a single serving of milk is about 1 cup.

Al estimated some measurements. Do Numbers 7 through 10.

7. Al estimated the width of the chalkboard. Which of these measures might have been his correct estimate?

Ⓐ 4 inches
Ⓑ 4 miles
Ⓒ 4 meters
Ⓓ 4 centimeters

8. Al found a small paper clip. He estimated its length. Which of these measures could be the length?

Ⓐ 1 yard
Ⓑ 1 meter
Ⓒ 1 foot
Ⓓ 1 inch

9. Al estimated the length of a new pencil. Which of these measures is most likely the correct length?

Ⓐ 6 centimeters
Ⓑ 6 inches
Ⓒ 6 yards
Ⓓ 6 meters

10. There are some games in Al's classroom. He measured the length of one edge of one of the dice. Which of these measures could be the length?

Ⓐ 1 yard
Ⓑ 1 foot
Ⓒ 1 meter
Ⓓ 1 centimeter

Read what Al wrote about his baby sister. Then do Numbers 11 through 14.

I have a baby sister. Her name is Anna. She is much smaller than I. I am tall enough to reach the doorknob. The doctor measured Anna. He told Dad that Anna is now 24 inches long. Anna is too small to play with me. But I do get to hold her. I sometimes get to feed her. She drinks a lot of milk! I try to help Mom and Dad with my sister. When it is time to go out, I help put on her shoes. I sometimes play "This Little Piggy" with her toes. When I do, she giggles. That must mean she likes my game!

11. Al plays "This Little Piggy" with Anna's toes. How long might one of her toes be?

Ⓐ 1 foot

Ⓑ 1 centimeter

Ⓒ 1 meter

Ⓓ 1 yard

12. Anna drinks about 8 cups of milk each day. How many pints of milk does she drink each day?

Ⓐ 4 pints

Ⓑ 16 pints

Ⓒ 2 pints

Ⓓ 12 pints

13. The doctor told Al's father that Anna is 24 inches long. How many feet is 24 inches?

Ⓐ 3 feet

Ⓑ 1 foot

Ⓒ 4 feet

Ⓓ 2 feet

14. Al likes to help with Anna's shoes. Which of these measures might be the length of her shoe?

Ⓐ 3 feet

Ⓑ 3 meters

Ⓒ 3 inches

Ⓓ 3 yards

- A test question about customary and metric measures may ask you to add or subtract measurements that are given in different units.
- A test question about customary and metric measures may ask you to use measurement facts.
- A test question about customary and metric measures may ask you to estimate the measurement of something.

Al measured the feet of everyone in his family. Read about his measurements. Then do Numbers 15 and 16.

Our Feet

First, I measured my feet. My feet are 8 inches long. Then I measured the feet of both of my sisters. My older sister's feet are 10 inches long. My baby sister's feet are 7 centimeters long. Next, I measured my mother's feet. Hers are 1 foot long. Finally, I measured my father's feet. His feet are 3 inches longer than 1 foot. He has the longest feet I have ever seen!

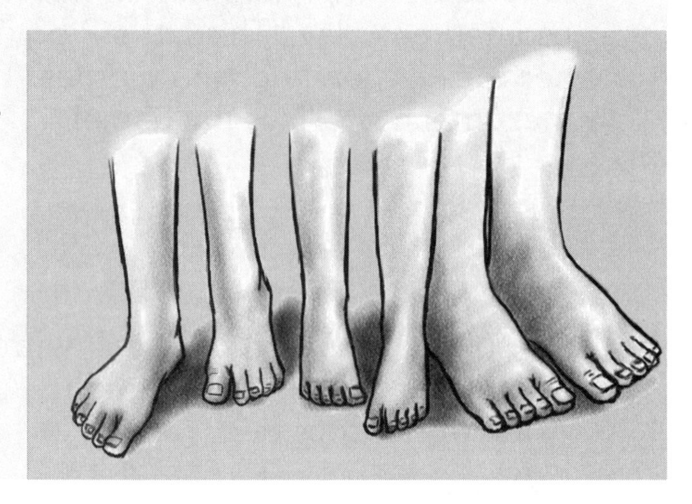

Converting Customary and Metric Measures

15. Al found how much longer his mom's foot is than his own foot.

What is the difference in the lengths?

Ⓐ 7 inches

Ⓑ 16 inches

Ⓒ 2 inches

Ⓓ 4 inches

Converting Customary and Metric Measures

16. Al's father has the longest feet in the family. They are 3 inches longer than 1 foot.

What is the length?

Ⓐ 15 inches

Ⓑ 4 inches

Ⓒ 14 inches

Ⓓ 9 inches

Al and his older sister have a fish tank. Read about their tank. Then do Numbers 17 and 18.

Our Fish Tank

Helen and I have a fish tank. In it, there are 5 gallons of water. My favorite fish is the angelfish. His name is Harvey. He is black and has long fins that look like wings. I think that is why he is called an angelfish. We also have 4 goldfish. Their names are Eenie, Meenie, Mynee, and Moe. Helen named them. Rocks and plants sit on the bottom of the tank. There is also a frog at the bottom. He is very shy. Whenever I go near the tank, he hides behind a rock. He is a little bit larger than the width of my thumb. That's why we named him Tiny.

Converting Customary and Metric Measures

17. The fish tank has 5 gallons of water in it.

How many quarts of water are in the tank?

- Ⓐ 9 quarts
- Ⓑ 20 quarts
- Ⓒ 1 quart
- Ⓓ 16 quarts

Converting Customary and Metric Measures

18. Tiny the frog is a little larger than the width of Al's thumb.

Which of these measures could be Tiny's length?

- Ⓐ 3 meters
- Ⓑ 3 feet
- Ⓒ 3 centimeters
- Ⓓ 3 yards

Strategy Nine USING ALGEBRA

PART ONE: Think About Algebra

WHAT DO YOU KNOW ABOUT NUMBER PATTERNS?

An IN and OUT machine has a number pattern that tells how it works.

Patterns are like rules.

The pattern for the IN and OUT machine is:

Add 3 to each IN number to get the OUT number.

IN	OUT
2	5
3	6
5	8
8	11

▶ Find the pattern and the missing number in each IN and OUT machine.

a.

IN	OUT
4	8
5	10
6	12
7	?

The pattern is: ______________________

______________________________________.

The missing OUT number is _____.

b.

IN	OUT
25	20
20	15
15	?
10	5

The pattern is: ______________________

______________________________________.

The missing OUT number is _____.

c.

IN	OUT
2	6
3	7
5	9
8	?

The pattern is: ______________________

______________________________________.

The missing OUT number is _____.

d.

IN	OUT
24	8
21	?
18	6
15	5

The pattern is: ______________________

______________________________________.

The missing OUT number is _____.

You just reviewed how to find number patterns in IN and OUT machines.

WHAT MORE DO YOU KNOW ABOUT NUMBER PATTERNS?

Patterns are like rules.

The numbers in a group may follow a pattern (or rule).

2, 4, 6, ____, 10

The pattern (or rule) for the group of numbers is:
Add 2 to each number to get the next number in the row.
The missing number in the row is 8.

▶ Find the pattern and the missing number in each row of numbers.

a. 20, 17, ____, 11, 8, 5

The pattern is: ______________________________.

The missing number in the row is ___________.

b. 12, ____, 36, 48, 60

The pattern is: ______________________________.

The missing number in the row is ___________.

c. 700, 600, ____, 400, 300

The pattern is: ______________________________.

The missing number in the row is ___________.

You just reviewed finding the missing number in a pattern.

Alone, write two patterns, each having a missing number. Give your patterns to your partner while you take your partner's patterns. Solve for the missing numbers. Together, discuss how you determined each missing number.

Evan made some IN and OUT machines for a math project. Each machine has a pattern (or rule) for how it works. Study the IN and OUT numbers. As you study, think about what the patterns might be.

1 + ? = 4	1 → IN OUT → 4	
2 + ? = 5	2 → IN OUT → 5	
3 + ? = 6	3 → IN OUT → 6	
4 + ? = 7	4 → IN OUT → 7	

Pattern: Add 3 to the IN number to get the OUT number.

10 − ? = 8	10 → IN OUT → 8
9 − ? = 7	9 → IN OUT → 7
8 − ? = 6	8 → IN OUT → 6
7 − ? = 5	7 → IN OUT → 5

Pattern: Subtract 2 from the IN number to get the OUT number.

Sometimes, you are asked for the missing number in a pattern. First, figure out the pattern. Then use the pattern to find the missing number.

10, 14, 18, ____, 26

$10 + 4 = 14$
$14 + 4 = 18$
$18 + 4 = \underline{22}$
$22 + 4 = 26$

Pattern: Add 4.
The missing number is 22.

50, 45, 40, ____, 30

$50 - 5 = 45$
$45 - 5 = 40$
$40 - 5 = \underline{35}$
$35 - 5 = 30$

Pattern: Subtract 5.
The missing number is 35.

You use **algebra** when you find patterns.

- Patterns are like rules.
- Use patterns to find missing numbers in a group.

Evan wrote these groups of numbers. Study each group of numbers. Think about the pattern (or rule) for each group. Then do Numbers 1 through 4.

Number Pattern	IN and OUT Machine
54, 48, 42, ____, 30	4 → IN OUT → 12 6 → IN OUT → 14 8 → IN OUT → 16 10 → IN OUT → ?

1. In the number pattern, what do you do to 54 to get an answer of 48?

   Ⓐ Add 10.

   Ⓑ Add 12.

   Ⓒ Subtract 6.

   Ⓓ Subtract 9.

2. What is the missing number in the number pattern?

   Ⓐ 48

   Ⓑ 36

   Ⓒ 32

   Ⓓ 33

3. Look at the IN and OUT machine. What do you do to 6 to get an answer of 14?

   Ⓐ Subtract 2.

   Ⓑ Add 2.

   Ⓒ Subtract 8.

   Ⓓ Add 8.

4. Evan put a 10 into the IN part of the machine. What was the OUT number?

   Ⓐ 2

   Ⓑ 18

   Ⓒ 12

   Ⓓ 8

Talk about your answers to questions 1–4.
Tell why you chose the answers you did.

PART THREE: Check Your Understanding

Remember: You use algebra when you find patterns.

- Patterns are like rules.
- Use patterns to find missing numbers in a group.

Solve this problem. As you work, ask yourself, "What can I do to one number to find the next number in the pattern?"

5. Evan wrote a pattern for his friend to solve. What number is missing?

210, 217, 224, ____, 238

Ⓐ 245

Ⓑ 217

Ⓒ 231

Ⓓ 203

Solve another problem. As you work, ask yourself, "What do I do to an IN number to get the correct OUT number?"

6. Evan put three numbers into the IN part of the machine. What is the last OUT number?

83 →	IN	OUT	→ 73
73 →	IN	OUT	→ 63
63 →	IN	OUT	→ ?

Ⓐ 53

Ⓑ 63

Ⓒ 93

Ⓓ 43

Look at the answer choices for each question.
Read why each answer choice is correct or not correct.

5. Evan wrote a pattern for his friend to solve. What number is missing?

210, 217, 224, ____, 238

Ⓐ 245

This answer is not correct because the pattern is to add 7. If you add 7 to 224, you get 231, not 245.

Ⓑ 217

This answer is not correct because the pattern is to add 7. If you add 7 to 224, you get 231, not 217.

● 231

This answer is correct because the pattern is to add 7.

210 + 7 = 217
217 + 7 = 224
224 + 7 = 231
231 + 7 = 238

Ⓓ 203

This answer is not correct because the pattern is to add 7. If you add 7 to 224, you get 231, not 203.

6. Evan put three numbers into the IN part of the machine. What is the last OUT number?

83 →	IN	OUT	→ 73
73 →	IN	OUT	→ 63
63 →	IN	OUT	→ ?

● 53

This answer is correct because, for this machine, you have to subtract 10 from the IN number to get the OUT number.

83 − 10 = 73
73 − 10 = 63
63 − 10 = 53

Ⓑ 63

This answer is not correct because if you subtract 10 from 63, you get 53.

Ⓒ 93

This answer is not correct because if you subtract 10 from 63, you get 53.

Ⓓ 43

This answer is not correct because if you subtract 10 from 63, you get 53.

PART FOUR: Learn More About Algebra

You can use algebra to help you solve problems.

- Read the problem. Think about the numbers you know.
  Decide if you would need to add or subtract to solve the problem.
  Choose the number sentence that solves the problem.

Evan and his team needed some new baseball equipment for Friday's game. Do Numbers 7 through 10.

7. Evan bought 1 glove and some baseballs. He bought 5 items in all. Which of these number sentences will help you figure out how many baseballs Evan bought?

   Ⓐ $5 + 1 = \square$

   Ⓑ $1 + \square = 1$

   Ⓒ $1 + \square = 5$

   Ⓓ $1 + \square = 12$

8. Evan looked at the shelf of baseball equipment shown above. Which of these number sentences will help you figure out how many more balls there are than bats?

   Ⓐ $12 - 5 = \square$

   Ⓑ $12 - \square = 5$

   Ⓒ $3 + 12 = \square$

   Ⓓ $12 - 3 = \square$

9. Evan bought 4 hats from the shelf above for his team. Which of these number sentences will help you figure out how many hats are left on the shelf?

   Ⓐ $7 - \square = 7$

   Ⓑ $7 - 4 = \square$

   Ⓒ $7 + 4 = \square$

   Ⓓ $7 + \square = 4$

10. There were 12 bats displayed in the store window. Evan wanted to know how many bats there were all together. Which of these number sentences should Evan use to figure out how many bats are in the store?

    Ⓐ $1 + \square = 12$

    Ⓑ $12 + 3 = \square$

    Ⓒ $2 + 3 + \square = 12$

    Ⓓ $12 - 3 = \square$

Read about the playground at Evan's school. Then do Numbers 11 through 14.

Evan and some of his friends like outdoor recess. They play on the swings, slide down the slide, and climb on the bars. They sometimes play games. Today, 4 children are on the bars, and 4 children are at the slide.

11. Evan and 3 of his friends played ball with some more children. In all, 8 children played ball. Which of these number sentences will help you figure out how many children joined in?

Ⓐ $8 + 3 = \square$

Ⓑ $4 + \square = 8$

Ⓒ $8 + 4 = \square$

Ⓓ $4 - 3 = \square$

12. Last week, 10 children waited in line for the slide. Then 4 of them ran to the swings. Which of these number sentences will help you figure out how many children were left in line?

Ⓐ $10 + 4 = \square$

Ⓑ $4 - \square = 10$

Ⓒ $4 + 10 = \square$

Ⓓ $10 - 4 = \square$

13. Which of these number sentences will help you figure out how many more children are on the bars than at the slide today?

Ⓐ $4 - \square = 0$

Ⓑ $4 + 4 = \square$

Ⓒ $4 - 4 = \square$

Ⓓ $0 + 4 = \square$

14. There are 8 swings, and 5 children are swinging. Which of these number sentences should Evan use to figure out how many swings are empty?

Ⓐ $8 - 5 = \square$

Ⓑ $8 + 5 = \square$

Ⓒ $5 - 5 = \square$

Ⓓ $13 - 5 = \square$

PART FIVE: Prepare for a Test

- A test question about algebra may ask for the missing number in a group.
- A test question about algebra may ask for an IN number or an OUT number.
- A test question about algebra may ask which number sentence can help solve a problem.

Evan's class went to the toy museum. Read about their visit. Then do Numbers 15 and 16.

The Toy Museum

Evan and his classmates enjoy building things with blocks. At the museum, they built tall buildings, spaceships, and houses. Evan decided to build a tower. He started out with 33 blocks.

Using Algebra

15. One child built a wall of red and blue blocks. Evan made this table to show how many blue blocks and red blocks are in each row of the wall.

Red Blocks	1	2	3	4
Blue Blocks	3	4	5	?

How many blue blocks are in the row that has 4 red blocks?

Ⓐ 8 blue blocks

Ⓑ 5 blue blocks

Ⓒ 6 blue blocks

Ⓓ 4 blue blocks

Using Algebra

16. When Evan finished his tower, he had 4 blocks left over.

Which of these number sentences should Evan use to figure out how many blocks are in his tower?

Ⓐ $33 + 4 = \square$

Ⓑ $\square - 33 = 4$

Ⓒ $33 - 4 = \square$

Ⓓ $33 - 33 = \square$

Evan lives near a park. Read about this park. Then do Numbers 17 and 18.

Ferndale Park

People like to visit Ferndale Park. They hike on the trails and play soccer and kickball on the fields. Some people have picnics in the summer. Others like to paddle canoes in the pond. One day, the people in Evan's neighborhood had a group picnic. There were 28 people at the picnic.

Using Algebra

17. More and more people visit Ferndale Park every day. The missing numbers in the group are the same as the number of people who visited the park on Saturday and Sunday.

Find the pattern. What are the two missing numbers?

32, 42, 52, ____, ____

Ⓐ 72 and 82

Ⓑ 12 and 22

Ⓒ 62 and 72

Ⓓ 52 and 62

Using Algebra

18. At Evan's neighborhood picnic, 9 people went canoeing.

Which of these number sentences should Evan use to figure out how many people did *not* go canoeing?

Ⓐ $28 - 9 = \square$

Ⓑ $19 - \square = 9$

Ⓒ $28 + 9 = \square$

Ⓓ $9 - \square = 28$

REVIEW

PART ONE: Read an Article

Read the article about people who walk to help others. Then do Numbers 1 through 6.

Every year on a Sunday in May, thousands of people gather in the city of Boston, Massachusetts, to walk. They do not walk just for exercise. They walk to raise money to help feed hungry people.

The Walk for Hunger started in 1969 in Quincy, Massachusetts. That first year, about 2,000 people each walked 29.6 miles. They raised $26,000. The walk now takes place in Boston and is only 20 miles long.

Each year, more and more people have taken part in the walk. In 2005, 35,000 people raised $3 million dollars for 400 emergency programs.

To raise money, each walker gets sponsors. The sponsors give a certain amount of money for each mile that the walker completes. For example, a sponsor may give 5¢, 10¢, 25¢, 50¢, $1.00, or more, for each mile.

On the day of the walk, walkers gather on a grassy park in the city of Boston. Each walker has to sign in. Walkers can sign in between 7 A.M. and 9 A.M.

Walkers do not have to walk the whole 20 miles. They can walk as little or as much as they are able. Many people find that the walk is easier and more fun if they walk with a friend. Some people walk with family members. Others walk with groups from their school, club, or town.

People talk and sing and have fun as they walk. They are given water along the way. Sometimes, clowns do tricks and bands play music for the walkers.

Around noon, some walkers stop and have a picnic lunch. Most people are happy to stop for a while and rest their feet. But they are also cheerful when they get going again. They know that every step they take helps to pay for meals for people in need.

Converting Time and Money

1. Katya's friend Alan agreed to pay 75¢ for every mile she walked.

   What group of coins totals 75¢?

   Ⓐ 1 quarter, 3 dimes, 4 nickels
   Ⓑ 2 quarters, 1 dime, 1 nickel
   Ⓒ 2 quarters, 1 dime, 2 nickels
   Ⓓ 4 dimes, 3 nickels, 5 pennies

Converting Customary and Metric Measures

4. Katya drank 8 cups of water during the walk.

   How many pints did she drink?

   2 cups = 1 pint

   Ⓐ 16 pints Ⓒ 2 pints
   Ⓑ 4 pints Ⓓ 6 pints

Converting Time and Money

2. Katya signed in early and started walking at the time shown. She walked for 4 hours and 15 minutes.

   At what time did she stop walking?

   Ⓐ 10:00 A.M. Ⓒ 9:15 A.M.
   Ⓑ 11:45 A.M. Ⓓ 10:30 A.M.

Using Algebra

5. From Katya's school, 68 students took part in the walk, and 16 walked the entire 20 miles.

   Which of these number sentences shows how many students did *not* walk 20 miles?

   Ⓐ 68 + 16 = □
   Ⓑ 16 − □ = 68
   Ⓒ 68 − 16 = □
   Ⓓ 20 − □ = 16

Converting Customary and Metric Measures

3. Katya walked 1 foot away from Mom. Ilia walked 5 inches closer to Mom than Katya.

   How far was Ilia from Mom?

   1 foot = 12 inches

   Ⓐ 2 inches Ⓒ 7 inches
   Ⓑ 6 inches Ⓓ 9 inches

Using Algebra

6. The missing number in the pattern is the same as the number of walkers who stopped for lunch.

   How many walkers stopped?

   340, 346, 352, ____, 364

   Ⓐ 334 walkers Ⓒ 346 walkers
   Ⓑ 380 walkers Ⓓ 358 walkers

PART TWO: Read a Story

Read a familiar story with a new ending.
Then do Numbers 7 through 12.

A Polite Girl After All

Everyone knows the story of Goldilocks and the Three Bears. At least, they think they do.

Goldilocks, a little girl with a head of golden curls, was walking in the woods near her house. Soon, she came upon an empty cottage. The cottage belonged to the Bear family. Papa, Mama, and Baby Bear had gone out for a walk while their supper cooled on the table.

Goldilocks pushed open the door to the cottage. She didn't knock. She just walked right in. Inside, she saw three rocking chairs in front of a cozy fire. Goldilocks sat in all three chairs. She rocked so hard in Baby Bear's chair that it fell apart.

Then Goldilocks saw bowls of porridge on the table. She ate from all three bowls. She liked Baby Bear's porridge the best and ate it all. She left the dirty spoon and bowl in the sink.

After that, Goldilocks climbed the stairs to the bedrooms to take a nap. She messed up all three beds before she fell asleep in Baby Bear's bed.

That's where Papa, Mama, and Baby Bear found her. When they woke her up to ask what she was doing in their house, she ran away.

You may think that the story ends here, but it doesn't.

Later that same day, Goldilocks returned to the cottage. This time, she knocked on the door. When Papa Bear opened the door, Goldilocks was standing there with her father.

"I'm sorry for coming into your house today without asking," Goldilocks said to the Bears. "I hope you'll forgive me. My father is going to help me fix Baby Bear's chair. Then I'll wash the sheets and remake the beds. I've also brought dinner to replace the one I ate. And after dinner, I'll wash the dishes."

The Bears forgave Goldilocks. From that day on, the Bears let her visit their cottage whenever she wished—as long as she knocked first.

Converting Time and Money

7. The clocks show what time Goldilocks fell asleep and what time the Bears woke her.

How long did Goldilocks sleep?

Ⓐ 1 hour
Ⓑ 2 hours and 15 minutes
Ⓒ 2 hours
Ⓓ 1 hour and 30 minutes

Converting Time and Money

8. Goldilocks used the coins shown to pay for wood to fix Baby Bear's chair.

How much did she spend?

Ⓐ 99¢ Ⓒ 94¢
Ⓑ 84¢ Ⓓ 89¢

Converting Customary and Metric Measures

9. A chair was as tall as Goldilocks.

Which of these measures might be the height of the chair?

Ⓐ 1 meter Ⓒ 1 foot
Ⓑ 1 inch Ⓓ 1 centimeter

Converting Customary and Metric Measures

10. Goldilocks learned that the Bears drink 3 gallons of milk a day.

How many quarts do they drink?

1 gallon = 4 quarts

Ⓐ 6 quarts
Ⓑ 9 quarts
Ⓒ 12 quarts
Ⓓ 1 quart

Using Algebra

11. Goldilocks saw a pattern in the number of cups and bowls on the shelves.

Cups	3	5	7	9
Bowls	6	8	10	?

How many bowls are on the shelf with 9 cups?

Ⓐ 7 bowls Ⓒ 12 bowls
Ⓑ 16 bowls Ⓓ 9 bowls

Using Algebra

12. Goldilocks brought 24 biscuits for dinner and had 5 left over.

What number sentence shows how many biscuits were eaten?

Ⓐ $24 - 5 = \square$ Ⓒ $24 + 4 = \square$
Ⓑ $24 + \square = 5$ Ⓓ $24 - 24 = \square$

PART ONE: Think About Plane Figures

WHAT DO YOU KNOW ABOUT PLANE FIGURES?

Plane figures are flat.

Circles, triangles, squares, and rectangles are plane figures.

▶ Draw each plane figure in the space provided.

a. Draw a square that is one inch on a side.

b. Draw a triangle that is one inch on a side.

c. Draw a rectangle that is 4 centimeters long and 2 centimeters wide.

d. Draw a small circle.

e. Draw a circle inside a square.

f. Draw a triangle inside a circle.

g. Draw a square inside a circle.

h. Draw two triangles inside a rectangle.

You just reviewed how to draw plane figures.

WHAT DO YOU KNOW ABOUT GEOMETRIC FIGURES?

A larger figure can be made up of smaller figures.

▶ Complete each sentence about each geometric figure below.

a.

The square is divided into 2 equal ___________.

b. 

The rectangle is divided into 6 equal ___________.

c. 

The chain is made up of 4 equal _________________.

d. 

The figure is made up of 6 equal ___________.

e. 

The figure is made up of 6 equal ___________.

You just worked with larger figures made up of smaller figures.

Together, draw a pattern of geometric figures. Use circles, triangles, squares, and rectangles. Label each shape.

Lori made a drawing of the shapes she saw in her room. Study the drawing. As you study, think about how many sides and faces each figure has.

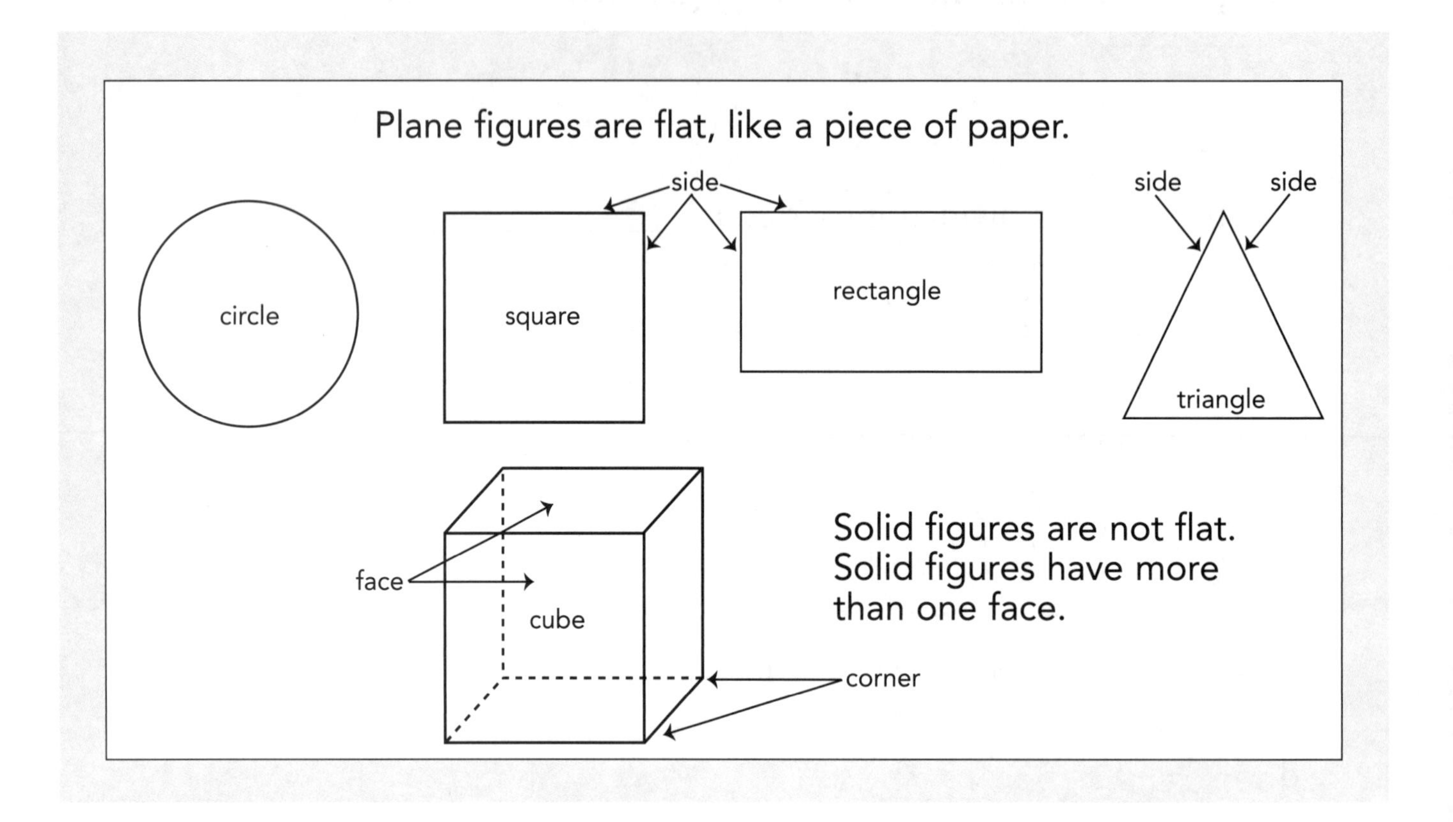

You use **geometry** to name figures.

- Plane figures are flat. Circles, squares, rectangles, and triangles are plane figures.
- Solid figures are not flat. Cubes are solid figures.

Lori visited the art museum. She bought a postcard of her favorite woven rug. Think about the figures on the rug. Then do Numbers 1 through 4.

1. Lori counted the white triangles on the rug. How many white triangles are there?

Ⓐ 14 white triangles

Ⓑ 10 white triangles

Ⓒ 6 white triangles

Ⓓ 3 white triangles

2. What is the name of the black figure in the middle of the rug?

Ⓐ square

Ⓑ circle

Ⓒ triangle

Ⓓ cube

3. Lori wants to buy a frame for the postcard she bought. What will the shape of the frame be?

Ⓐ square

Ⓑ triangle

Ⓒ rectangle

Ⓓ circle

4. At the gift shop, Lori saw a box shaped like a cube. How many faces were on the box?

Ⓐ 2 faces

Ⓑ 6 faces

Ⓒ 4 faces

Ⓓ 8 faces

Talk about your answers to questions 1–4.
Tell why you chose the answers you did.

PART THREE: Check Your Understanding

Remember: You use geometry to name figures.

- Plane figures are flat. Circles, squares, rectangles, and triangles are plane figures.
- Solid figures are not flat. Cubes are solid figures.

Solve this problem. As you work, ask yourself, "What faces on the boxes can be seen? What faces are hidden?"

5. Lori has a stack of wooden blocks on the floor. She walks around the blocks and counts only the faces she can see. How many faces can Lori see?

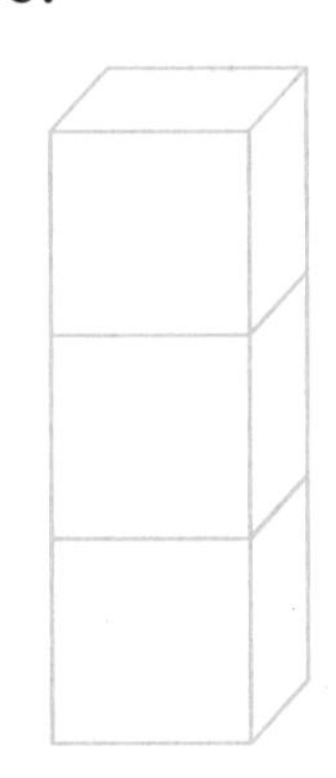

Ⓐ 18 faces
Ⓑ 12 faces
Ⓒ 13 faces
Ⓓ 14 faces

Solve another problem. As you work, ask yourself, "How many sides does a rectangle have? What does a circle look like?"

6. The floor of Lori's room is shaped like a rectangle. In the middle of her floor is a rug shaped like a circle. What picture shows the look of Lori's floor?

Ⓐ 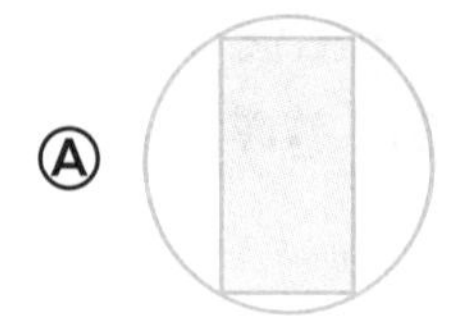

Ⓑ 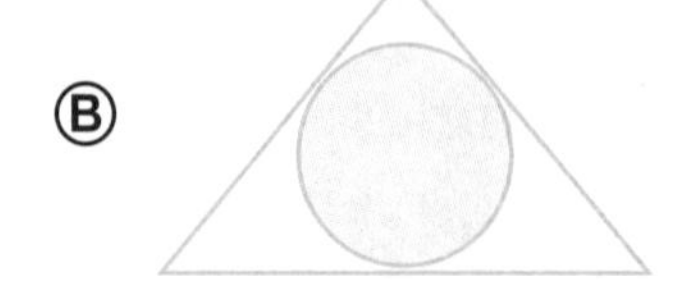

Ⓒ 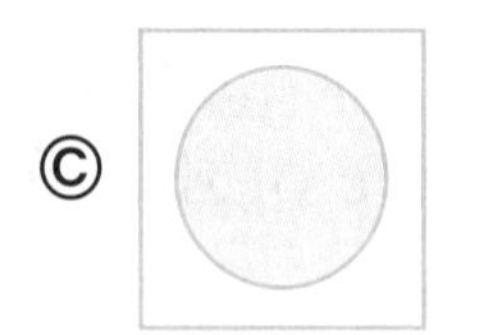

Ⓓ 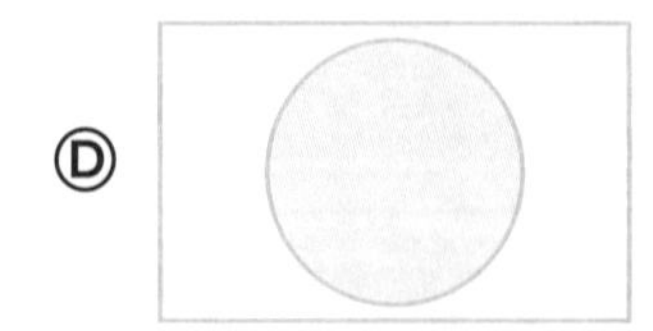

**Look at the answer choices for each question.
Read why each answer choice is correct or not correct.**

5. Lori has a stack of wooden blocks on the floor. She walks around the blocks and counts only the faces she can see. How many faces can Lori see?

Ⓐ 18 faces

This answer is not correct because this would mean that all faces of all the boxes could be seen. Some faces are hidden.

Ⓑ 12 faces

This answer is not correct because 12 is the number of faces that can be seen on all 4 sides. There is 1 more face on the top.

● 13 faces

This answer is correct because 13 is the number of faces that can be seen on all 4 sides plus the 1 face on the top: $3 + 3 + 3 + 3 + 1 = 13$.

Ⓓ 14 faces

This answer is not correct because this would mean that the faces of all 4 sides plus the top and the bottom could be seen. The bottom is hidden.

6. The floor of Lori's room is shaped like a rectangle. In the middle of her floor is a rug shaped like a circle. What picture shows the look of Lori's floor?

Ⓐ 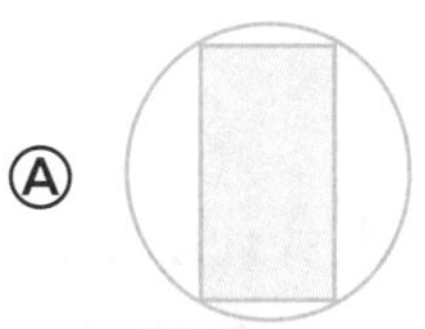

This answer is not correct because it shows a rectangle inside a circle. The circle should be inside the rectangle.

Ⓑ 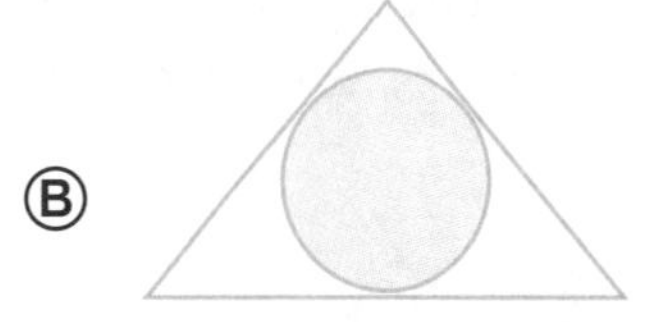

This answer is not correct because it shows a circle inside a triangle. The circle should be inside a rectangle.

Ⓒ 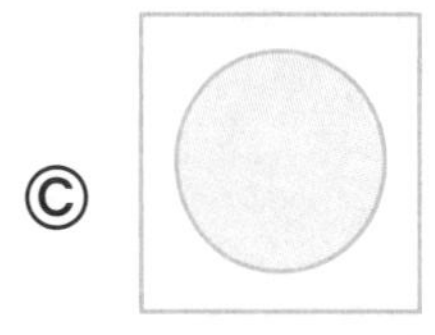

This answer is not correct because it shows a circle inside a square. The circle should be inside a rectangle.

● 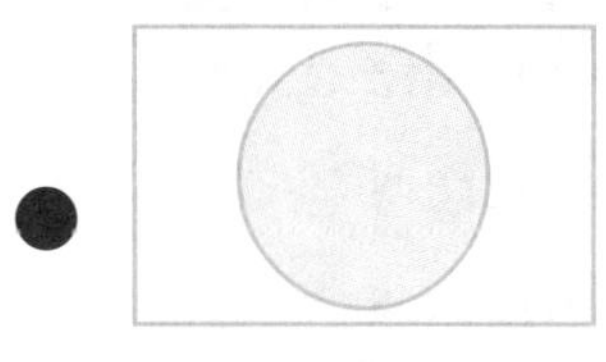

This answer is correct because it shows a circle inside a rectangle.

You use geometry to picture and to measure figures.

- When some figures are folded, the two halves match.
- A larger figure may be pictured as made up of smaller figures.
- The distance around a figure can be measured.

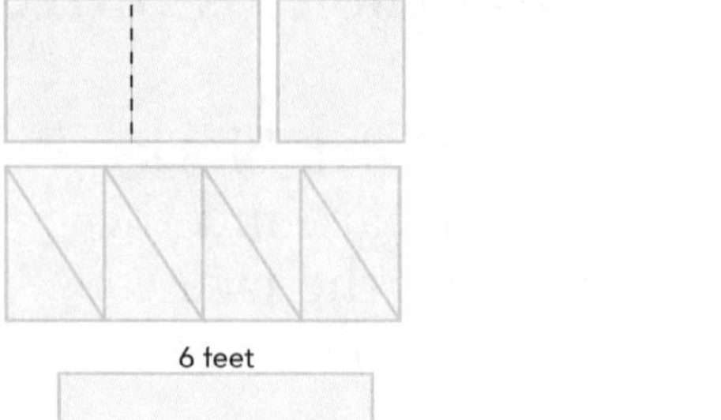

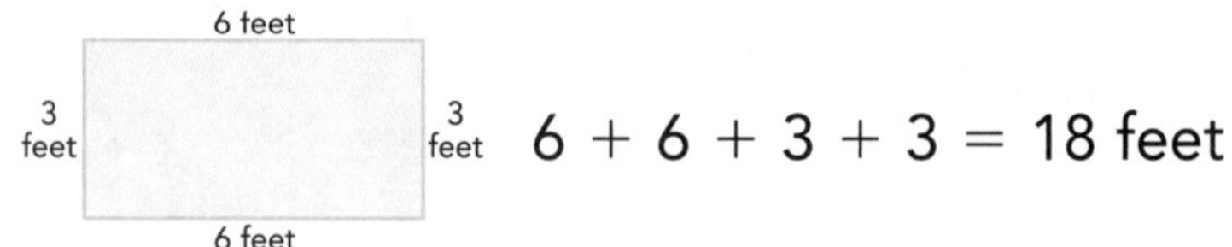

Lori has a box filled with craft items. On rainy days, she likes to make different things. Do Numbers 7 through 10.

7. This picture shows a place mat that Lori is making. Each triangle piece will be a different color. How many colors will Lori use?

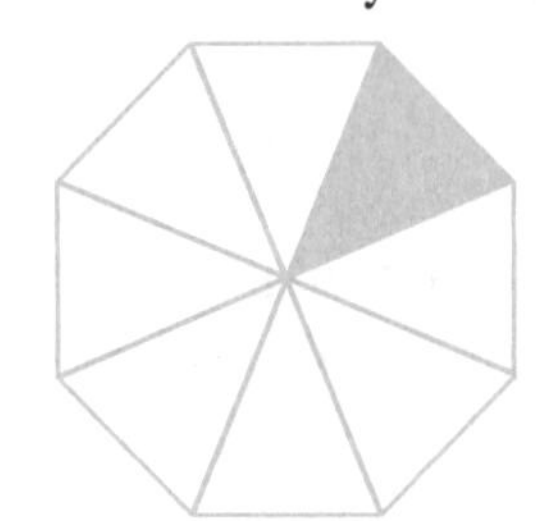

Ⓐ 6 colors
Ⓑ 8 colors
Ⓒ 4 colors
Ⓓ 10 colors

8. Lori will put lace all around the place mat. Each side of the place mat is 10 centimeters long. How much lace will she use?

Ⓐ 20 centimeters
Ⓑ 60 centimeters
Ⓒ 10 centimeters
Ⓓ 80 centimeters

9. Lori made a birthday card. She folded a piece of paper and then cut out a triangle.

Which of these shows what the paper will look like when it is unfolded?

Ⓐ 

Ⓒ  

Ⓑ 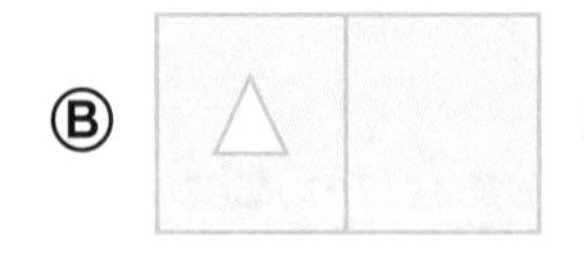

Ⓓ 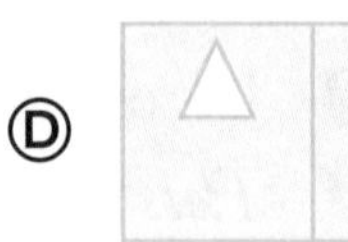

10. Lori glued a strip of material around all the sides of her craft box. The box is 12 inches long and 8 inches wide. How long is the strip of material?

Ⓐ 20 inches
Ⓑ 24 inches
Ⓒ 16 inches
Ⓓ 40 inches

Lori uses pattern shapes in math class. Look at the shapes she uses. Then do Numbers 11 through 14.

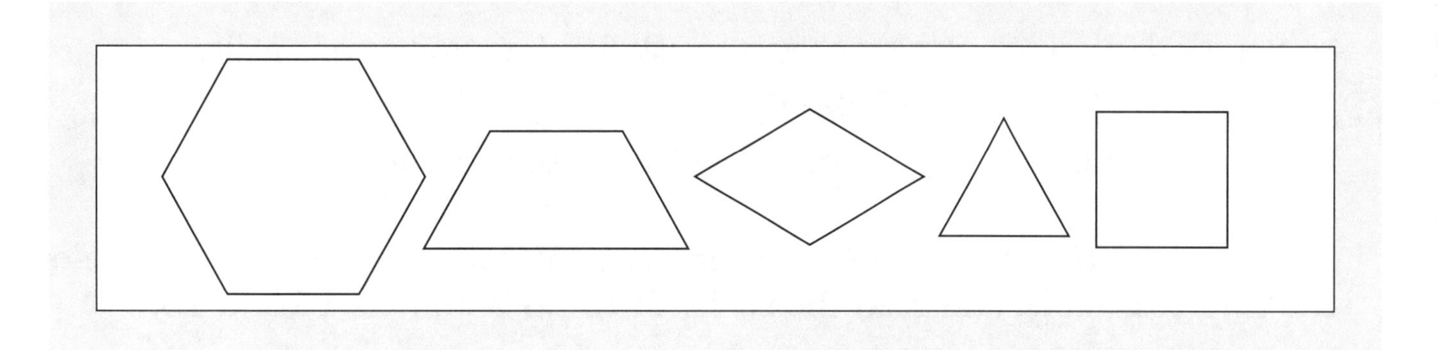

11. Lori used triangles to cover these figures. How many triangles did she use to cover both figures?

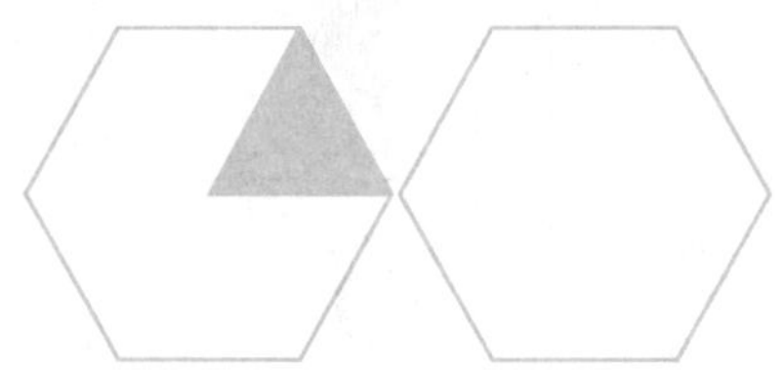

Ⓐ 6 triangles
Ⓑ 12 triangles
Ⓒ 15 triangles
Ⓓ 18 triangles

12. Each side of the pattern shapes shown below is 1 inch long. What is the total distance around this figure that Lori made?

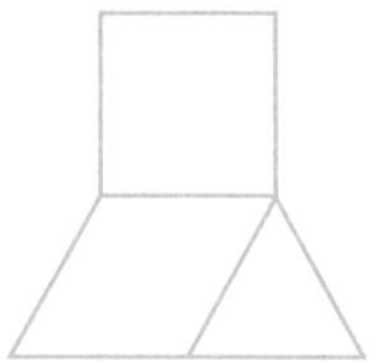

Ⓐ 5 inches
Ⓑ 9 inches
Ⓒ 7 inches
Ⓓ 8 inches

13. Lori folded this shape on the dotted line. What does the folded figure look like?

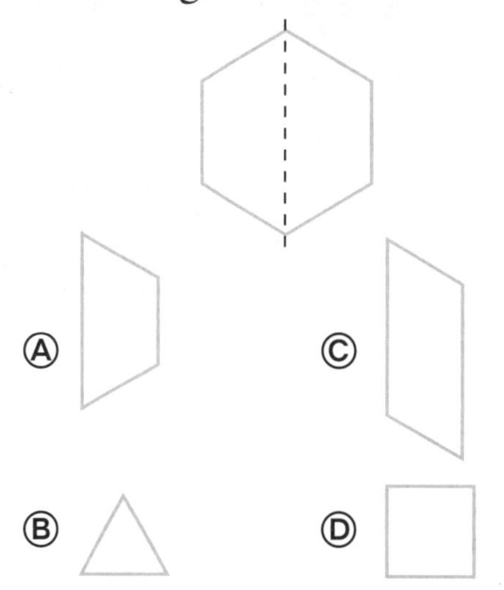

14. Lori drew this figure. Then she placed same-sized triangle pieces over it. How many triangles did she use to cover the figure?

Ⓐ 4 triangles
Ⓑ 6 triangles
Ⓒ 5 triangles
Ⓓ 8 triangles

- A test question about geometry may ask for the name of a figure.
- A test question about geometry may ask for the number of faces that can be seen on a solid figure.
- A test question about geometry may ask for the number of smaller figures that cover a larger figure.
- A test question about geometry may ask for the distance around a figure.

Lori played a ring-toss game at the amusement park. Look at the picture of the ring-toss table, and read the rules. Then do Numbers 15 and 16.

Rules for the Ring-Toss Game

- Throw the ring. If the ring lands around a prize, you win it.
- The ring must land flat on the table to win.

Good luck!

3 tosses for $1.00

Using Geometry

15. Lori saw that there are 4 parts to the ring-toss table. Each part is shaped like a figure.

What are the numbers and names of the 4 figures?

Ⓐ 1 rectangle, 2 squares, 1 triangle

Ⓑ 4 rectangles

Ⓒ 1 rectangle, 1 square, 2 triangles

Ⓓ 2 squares and 2 triangles

Using Geometry

16. This is the ring used in the game.

What is the name of this figure?

Ⓐ circle

Ⓑ square

Ⓒ cube

Ⓓ triangle

Lori's family is having a fence put up in their yard. Look at the picture of the fence posts. Then do Numbers 17 and 18.

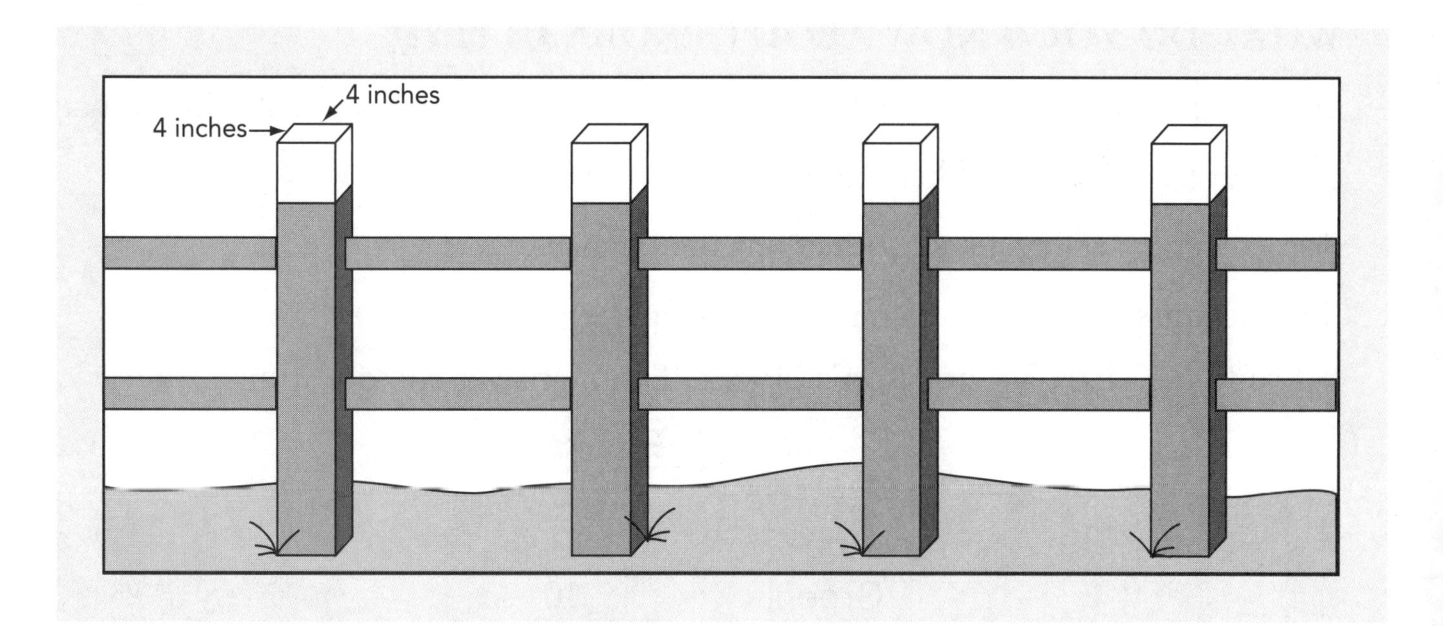

Using Geometry

17. The workers put a cube on the top of 4 fence posts. Lori walked around the posts and counted the number of faces she could see on all the cubes.

    How many faces could she see?

    Ⓐ 16 faces
    Ⓑ 20 faces
    Ⓒ 24 faces
    Ⓓ 12 faces

Using Geometry

18. Lori wanted to measure the distance around 1 cube. She tied a string around the cube. Then she measured the string.

    What was the length of the string?

    Ⓐ 4 inches
    Ⓑ 8 inches
    Ⓒ 12 inches
    Ⓓ 16 inches

DETERMINING PROBABILITY AND AVERAGES

PART ONE: Think About Probability and Averages

WHAT DO YOU KNOW ABOUT PROBABILITY?

You use probability to find what the chance is that a certain event will happen.

Some events are more likely to happen than others.

Some events are less likely to happen than others.

Some events have an equal chance of happening.

▶ Study the chart about Jason's blocks. Then, answer the questions.

Jason's Blocks	
Red	6
Blue	8
Green	10
Yellow	6

Jason put his blocks into a bag. He reaches into the bag, without looking, and chooses one block.

a. How many blocks did Jason place in the bag? ________

b. What color of block is Jason most likely to choose? ________

c. What color of block is Jason least likely to choose? ________ or ________

d. What two colors of blocks is Jason equally like to choose?

________ and ________.

You just reviewed probability.

WHAT DO YOU KNOW ABOUT AVERAGES?

You can find the average of a group of numbers.

First, add to find the total number of items in all the groups.

Next, put the total number of items into equal groups.

The number of equal groups should be the same as the number of groups that you had at the beginning.

▶ Study the chart of soda cans collected by second graders in five days.

Soda Cans	
Day	**Number of Cans**
Monday	10
Tuesday	6
Wednesday	10
Thursday	8
Friday	16

a. What is the total number of soda cans collected? ________

b. What is the total number of days it took to collect all the cans?______

c. Divide the total number of cans by the total number of days. This will give you the average. The average number of cans for each day is ________.

You just reviewed how to find an average.

Together, write one problem about finding an average. Write three questions about the problem. Answer the questions and discuss the results.

PART TWO: Learn About Probability

Roberto likes jelly beans. With his eyes closed, he will take a jelly bean from a plate. Study the chart of the colors and numbers of the jelly beans. As you study, think about what color Roberto will most likely take.

This chart shows the colors and the number of jelly beans of each color.

Color	Number
White	3
Orange	4
Pink	4
Red	8

The greatest number of jelly beans is 8 red jelly beans. Roberto is most likely to pick a red jelly bean.

The least number of jelly beans is 3 white jelly beans. Roberto is least likely to pick a white jelly bean.

You can also compare the chances of picking two different colors.

There are an equal number of orange and pink jelly beans. There are 4 of each. So, Roberto has an equal chance of picking an orange jelly bean as he has of picking a pink jelly bean.

There are more orange jelly beans than white ones. Roberto is more likely to pick an orange jelly bean than a white one.

There are fewer pink jelly beans than red ones. Roberto is less likely to pick a pink jelly bean than a red one.

You use **probability** to find what the chance is that a certain event will happen.

- Some events are more likely to happen than others.
- Some events are less likely to happen than others.
- Some events have an equal chance of happening. They are equally likely.

Roberto made a list of the shirts that he has in his closet. Think about how many shirts he has of each color. Then do Numbers 1 through 4.

2 White Shirts
3 Gray Shirts
6 Blue Shirts
3 Red Shirts
1 Black Shirt

1. With his eyes closed, Roberto takes a shirt from his closet. What color is he least likely to pick?

   Ⓐ white
   Ⓑ gray
   Ⓒ black
   Ⓓ blue

2. What color shirt is Roberto more likely to pick than a gray shirt?

   Ⓐ white
   Ⓑ red
   Ⓒ blue
   Ⓓ black

3. All of Roberto's shirts are now in the laundry basket. Roberto's dad reaches in and pulls out a shirt. What color will he most likely pick?

   Ⓐ red
   Ⓑ blue
   Ⓒ gray
   Ⓓ black

4. Which of these two colors of shirts is Roberto's dad equally likely to pick?

   Ⓐ gray and red
   Ⓑ red and blue
   Ⓒ black and blue
   Ⓓ red and white

Talk about your answers to questions 1–4.
Tell why you chose the answers you did.

PART THREE: Check Your Understanding

Remember: You use probability to find what the chance is that a certain thing will happen.

- Some events are more likely to happen than others.
- Some events are less likely to happen than others.
- Some events have an equal chance of happening. They are equally likely.

Solve this problem. As you work, ask yourself, "What item has the greatest number?"

5. In a basket, Roberto has 3 red blocks, 5 green blocks, 2 white blocks, and 2 yellow blocks. If he closes his eyes and picks a block, what color will he most likely pick?

   Ⓐ yellow
   Ⓑ red
   Ⓒ green
   Ⓓ white

Solve another problem. As you work, ask yourself, "What item has the least number?"

6. Roberto uses this spinner to play a game. What number is the spinner least likely to stop on?

   Ⓐ 1
   Ⓑ 2
   Ⓒ 3
   Ⓓ 4

Look at the answer choices for each question.
Read why each answer choice is correct or not correct.

5. In a basket, Roberto has 3 red blocks, 5 green blocks, 2 white blocks, and 2 yellow blocks. If he closes his eyes and picks a block, what color will he most likely pick?

Ⓐ yellow

This answer is not correct because there are only 2 yellow blocks. There are more red blocks and more green blocks than yellow blocks.

Ⓑ red

This answer is not correct because there are 3 red blocks and 5 green blocks. There are more green blocks than red blocks.

● green

This answer is correct because there are 5 green blocks. There are more green blocks than any other color of block. So, Roberto is most likely to pick a green block.

Ⓓ white

This answer is not correct because there are only 2 white blocks. There are more red blocks and green blocks than white blocks.

6. Roberto uses this spinner to play a game. What number is the spinner least likely to stop on?

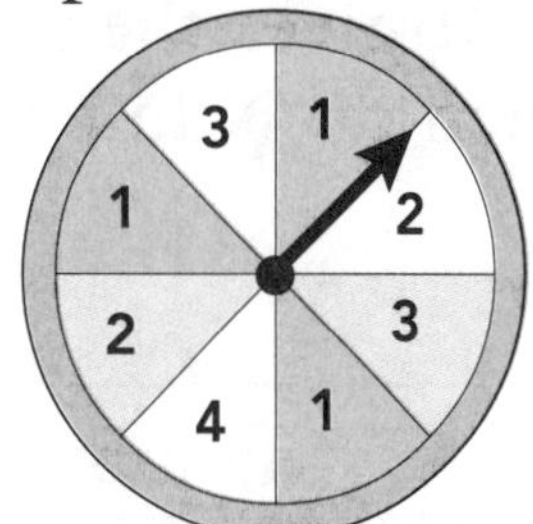

Ⓐ 1

This answer is not correct because there are more 1s than any other number, so the spinner is most *likely to stop on a 1.*

Ⓑ 2

This answer is not correct because there are more 2s than 4s, so the spinner is more *likely to stop on a 2 than on the 4.*

Ⓒ 3

This answer is not correct because there are more 3s than 4s, so the spinner is more *likely to stop on a 3 than on the 4.*

● 4

This answer is correct because 4 is on only 1 part of the spinner. All the other numbers have more than 1 part of the spinner. So, the spinner is least *likely to stop on the 4.*

PART FOUR: Learn About Averages

▶ You can find the **average** of a group of numbers.

Find the total number of items in all the groups. Put the total number of items into equal groups. Make the same number of groups that there were at the start.

Roberto reads in his science book every day at school. Study the chart. It gives the number of pages he read each day for five days. Then do Numbers 7 through 10.

Day	Number of Pages
Monday	2
Tuesday	3
Wednesday	4
Thursday	2
Friday	4

7. What is the total number of pages that Roberto read on these days?

Ⓐ 15 pages Ⓒ 5 pages

Ⓑ 10 pages Ⓓ 12 pages

8. What picture shows the total number of pages in equal groups?

Ⓐ 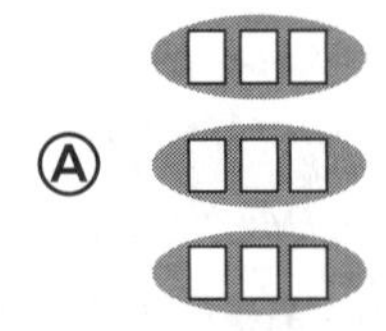

Ⓒ 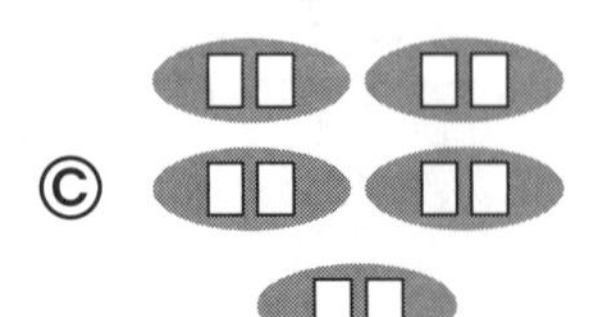

Ⓑ 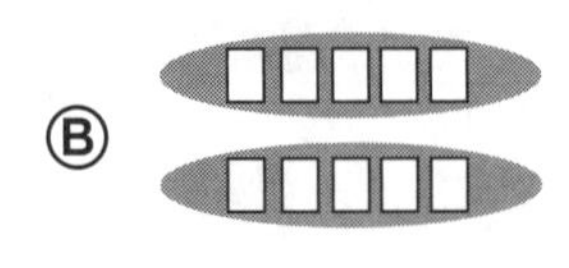

Ⓓ 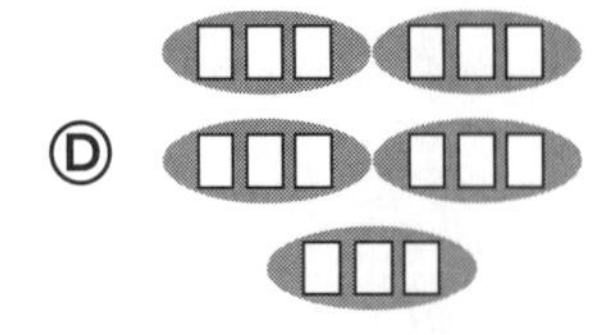

9. How many groups (or days) are shown in the chart?

Ⓐ 3 groups

Ⓑ 4 groups

Ⓒ 5 groups

Ⓓ 2 groups

10. The average is the number of items in each equal group. What is the average number of pages Roberto read each day?

Ⓐ 15 pages Ⓒ 5 pages

Ⓑ 10 pages Ⓓ 3 pages

Read about field day at Roberto's school. Then do Numbers 11 through 14.

Roberto played 4 different games at field day. At the ring toss, he won 2 tickets. Roberto won 3 tickets at the baseball throw and 3 tickets at the basketball hoop. Finally, he won 4 tickets at the dart game.

11. What is the total number of tickets that Roberto won?

Ⓐ 12 tickets

Ⓑ 14 tickets

Ⓒ 4 tickets

Ⓓ 16 tickets

12. Roberto played 4 different games. What is the average number of tickets he won at each game?

Ⓐ 12 tickets

Ⓑ 5 tickets

Ⓒ 3 tickets

Ⓓ 2 tickets

13. There were 3 people in the pie-eating contest. The number of pieces of pie that each person ate is shown in the box. What is the average number of pieces of pie that each person ate?

4	8	6

Ⓐ 4 pieces

Ⓑ 6 pieces

Ⓒ 18 pieces

Ⓓ 5 pieces

14. Roberto and his brothers brought home prizes from field day. What is the average number of prizes that each boy won?

Roberto won 4 prizes.
Pablo won 3 prizes.
Tomás won 2 prizes.

Ⓐ 2 prizes

Ⓑ 3 prizes

Ⓒ 5 prizes

Ⓓ 9 prizes

PART FIVE: Prepare for a Test

- A test question about probability may ask what event is most likely to happen.
- A test question about probability may ask what events are equally likely to happen.
- A test question about probability may ask what event is least likely to happen.
- A test question about averages may ask for an average of a group of items.

Roberto does a card trick with the number cards shown. Study the cards. Then do Numbers 15 and 16.

Determining Probability and Averages

15. Roberto holds the cards so that Tomás cannot see the numbers. Roberto asks Tomás to pick a card.

What number is Tomás most likely to pick?

Ⓐ 3
Ⓑ 1
Ⓒ 2
Ⓓ 5

Determining Probability and Averages

16. Roberto places each card on the table so that he cannot see the numbers. He then picks up a card.

What two numbers is Roberto equally likely to pick?

Ⓐ 3 and 4
Ⓑ 1 and 2
Ⓒ 2 and 3
Ⓓ 4 and 5

Roberto's school is collecting money for the food bank. Each day, students put nickels and dimes into a box. Study the chart that shows how many coins Roberto gave in 5 days. Then do Numbers 17 and 18.

Day	Nickels	Dimes
Monday	4	2
Tuesday	3	1
Wednesday	2	3
Thursday	2	3
Friday	4	1

Give your nickels and dimes to the food bank.

Determining Probability and Averages

17. What is the average number of dimes that Roberto gave each day?

Ⓐ 2 dimes

Ⓑ 3 dimes

Ⓒ 9 dimes

Ⓓ 10 dimes

Determining Probability and Averages

18. What is the average number of nickels that Roberto gave each day?

Ⓐ 5 nickels

Ⓑ 10 nickels

Ⓒ 3 nickels

Ⓓ 15 nickels

Strategy Twelve

INTERPRETING GRAPHS AND CHARTS

PART ONE: Think About Graphs and Charts

WHAT DO YOU KNOW ABOUT GRAPHS?

A bar graph uses bars and numbers to show how many.

A pictograph uses pictures to show how many.

▶ Study the bar graph and the pictograph. They both show the same information about ball games that a group of second graders enjoy.

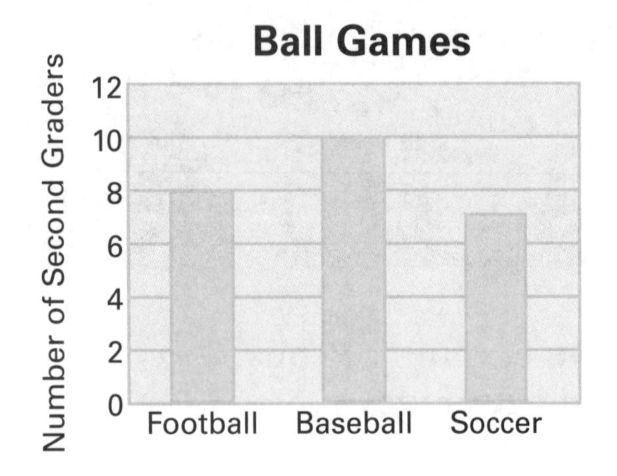

Ball Games	Number
Football	•••• ••••
Baseball	••••• •••••
Soccer	••• ••••

• = 1 second grader

a. How many second graders are shown on each graph? _______

b. How many ball games are shown on each graph? ________

c. How many second graders chose baseball? _______

d. How many second graders chose football? _______

e. How many second graders chose soccer? _______

You just reviewed how a bar graph and a pictograph can show the same information.

WHAT DO YOU KNOW ABOUT CHARTS?

A calendar is a kind of chart.

▶ Complete the calendar by adding the missing dates. Then answer the questions about the calendar.

DECEMBER						
Sunday	Monday	Tuesday	Wednesday	Thursday	Friday	Saturday
		1				
				31		

a. What date is the first Monday of December? __________

b. What date is the third Saturday of December? __________

c. What date is the fourth Wednesday of December? __________

d. On what day is December 15? _________

e. On what day is December 12? _________

f. How many Mondays are in December? _______

g. How many Wednesdays are in December? _________

You just reviewed how to read a calendar.

Together, create a June calendar. Write the days and the dates. Then, alone, write four questions about the June calendar. Trade questions with your partner and answer each other's questions. Then discuss the answers.

PART TWO: Learn About Graphs and Charts

Chantel loves skateboarding, skating, and riding her bicycle. She made some graphs and charts to show how 10 of her friends also like to do these things. Study the graphs and charts. As you study, think about the information that is given.

Count the dots. How many friends like to do each activity?

Skateboarding = 5 friends
Skating = 7 friends
Bicycling = 8 friends

Activity	Number
Skateboarding	● ● ● ● ●
Skating	● ● ● ● ● ● ●
Bicycling	● ● ● ● ● ● ● ●

● = 1 friend

You can show this same information in a bar graph.

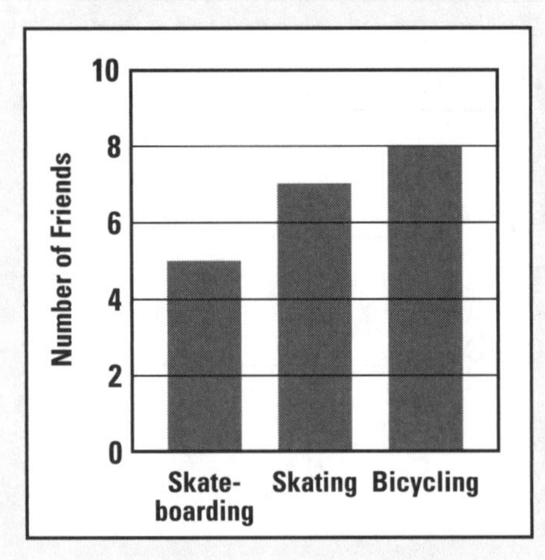

Look for the number that lines up with the top of each bar. The skateboarding bar is halfway between 4 and 6, so it stands for 5. The skateboarding bar stands for 7. The bicycling bar is at 8.

A Venn diagram can show how many go with one item only or with both items.

The 6 is in the bicycling circle only. So, 6 friends like bicycling only.

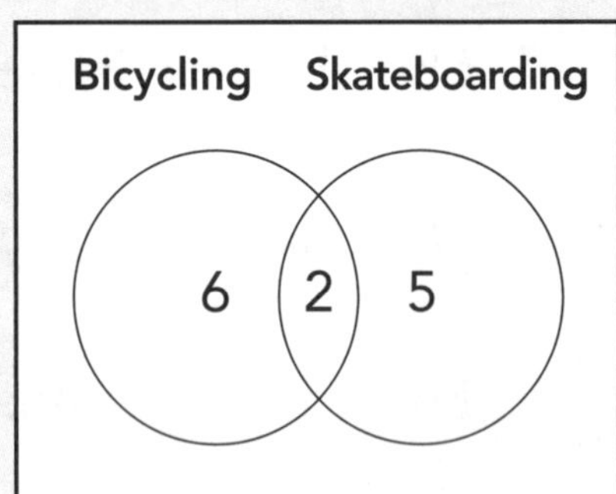

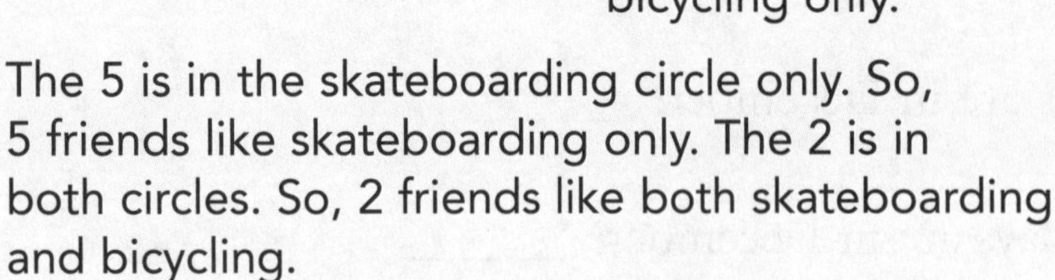

The 5 is in the skateboarding circle only. So, 5 friends like skateboarding only. The 2 is in both circles. So, 2 friends like both skateboarding and bicycling.

You use **graphs and charts** to show information.

- A pictograph uses pictures to show how many.
- A bar graph uses bars and numbers to show how many.
- A Venn diagram shows how many belong with one item only or with both items.

Chantel asked her friends what time they get up in the morning and what time they go to bed at night. Think about how many friends the graphs show for each time. Then do Numbers 1 through 4.

Get-Up Times (A.M.)

Time	Number
6:30	☺
6:45	☺ ☺
7:00	☺ ☺ ☺ ☺ ☺
7:15	☺ ☺

☺ = 1 friend

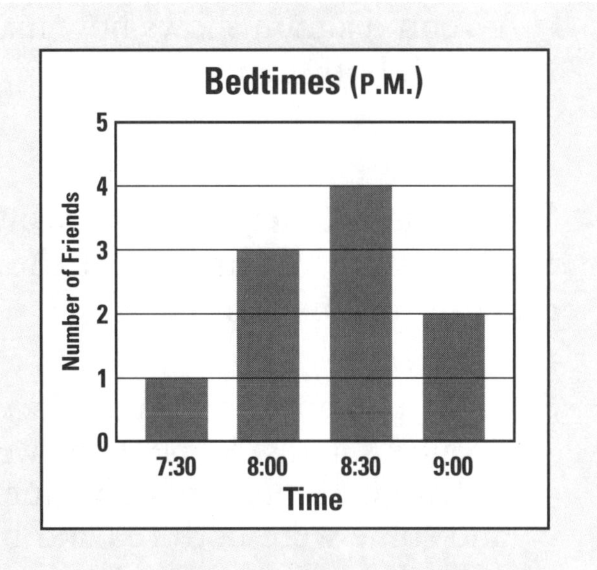

1. How many of Chantel's friends get up at 6:45 A.M.?
   Ⓐ 1 friend
   Ⓑ 2 friends
   Ⓒ 3 friends
   Ⓓ 4 friends

2. At what time do the greatest number of friends get up?
   Ⓐ 6:30 A.M.
   Ⓑ 6:45 A.M.
   Ⓒ 7:00 A.M.
   Ⓓ 7:15 A.M.

3. How many of Chantel's friends go to bed at 7:30 P.M.?
   Ⓐ 1 friend
   Ⓑ 2 friends
   Ⓒ 3 friends
   Ⓓ 5 friends

4. At what time do 3 of Chantel's friends go to bed?
   Ⓐ 7:30 P.M.
   Ⓑ 8:00 P.M.
   Ⓒ 8:30 P.M.
   Ⓓ 9:00 P.M.

Talk about your answers to questions 1–4.
Tell why you chose the answers you did.

PART THREE: Check Your Understanding

Remember: You use graphs and charts to show information.

- A pictograph uses pictures to show how many.
- A bar graph uses bars and numbers to show how many.
- A Venn diagram shows how many belong with one item only or with both items.

Solve this problem. As you work, ask yourself, "What part of the diagram means 'both'?"

5. Chantel made a diagram of a group of items. Some items were only red. Some were only blue, and some were both red and blue. How many items were both red and blue?

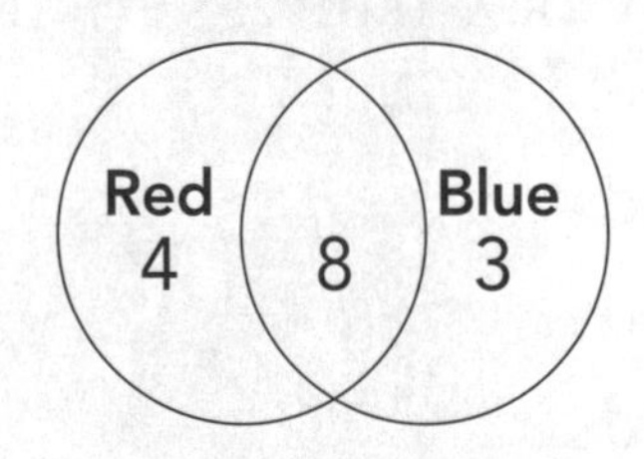

Ⓐ 3 items
Ⓑ 4 items
Ⓒ 8 items
Ⓓ 12 items

Solve another problem. As you work, ask yourself, "What picture matches the number given for each color and size?"

6. Chantel put 3 large white rocks and 2 small black rocks into her rock box. What box shows the rocks?

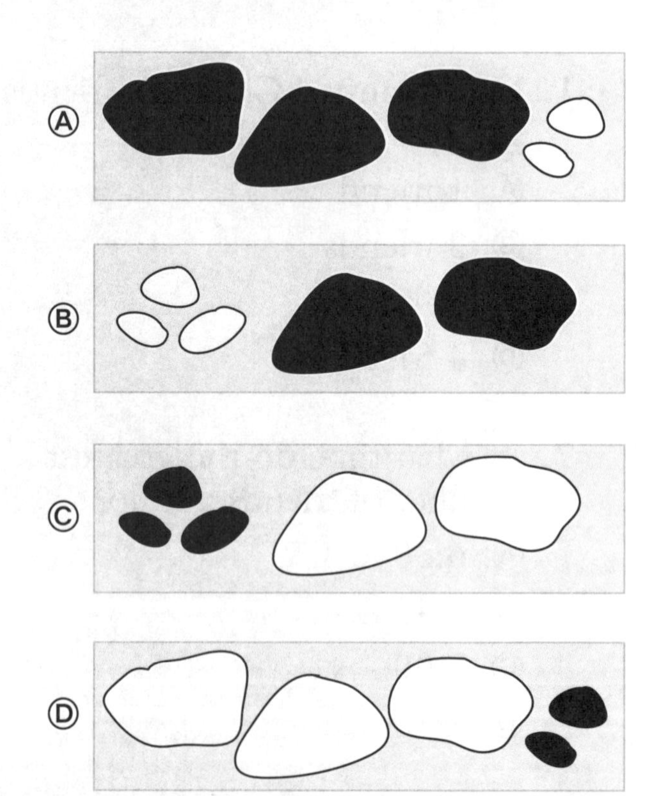

Look at the answer choices for each question.
Read why each answer choice is correct or not correct.

5. Chantel made a diagram of a group of items. Some items were only red. Some were only blue, and some were both red and blue. How many items were both red and blue?

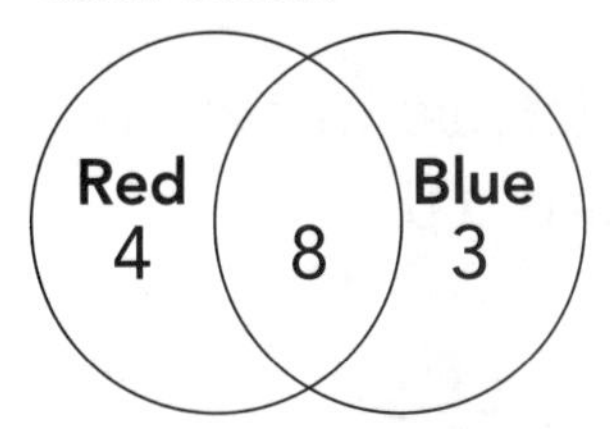

Ⓐ 3 items

This answer is not correct because the number 3 is in the blue circle only.

Ⓑ 4 items

This answer is not correct because the number 4 is in the red circle only.

● 8 items

This answer is correct because the number 8 is in both the red circle and the blue circle.

Ⓓ 12 items

This answer is not correct because it is the number of items that are only red (4) and the number of items (8) that are both red and blue.

6. Chantel put 3 large white rocks and 2 small black rocks into her rock box. What box shows the rocks?

Ⓐ 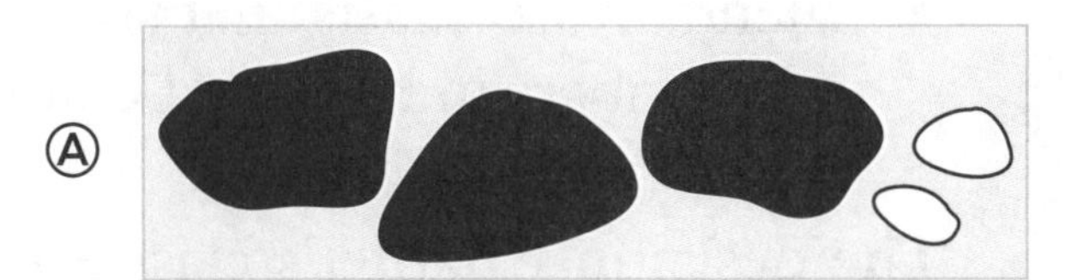

This answer is not correct because it shows 3 large black rocks and 2 small white rocks.

Ⓑ 

This answer is not correct because it shows 3 small white rocks and 2 large black rocks.

Ⓒ 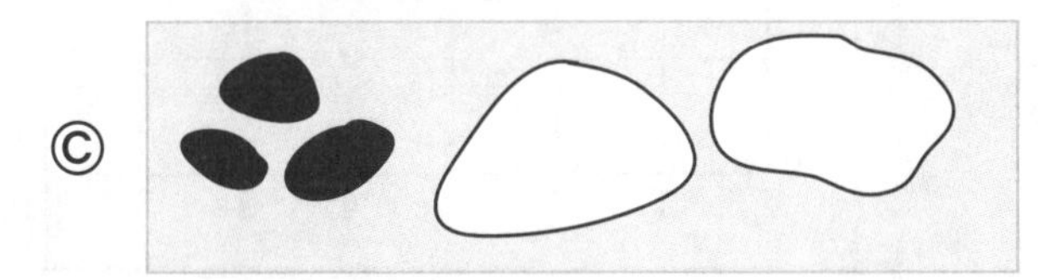

This answer is not correct because it shows 3 small black rocks and 2 large white rocks.

● 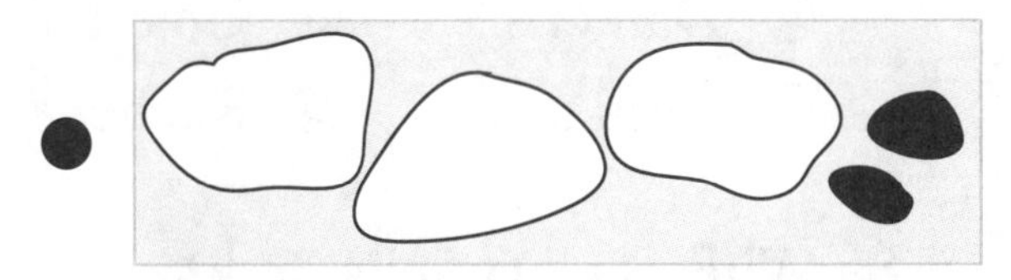

This answer is correct because it shows 3 large white rocks and 2 small black rocks.

PART FOUR: Learn More About Graphs and Charts

You use **graphs** and **charts** to show information in rows and columns.

- Some graphs, or grids, have points that can be described by location.

  In this grid, the star is over 1 and up 3. The box is over 2 and up 1.

- Calendars are charts that show days and numbers in rows and columns.

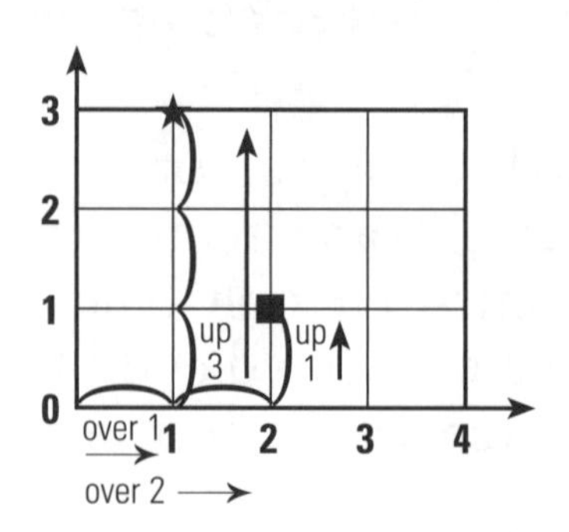

The desks in Chantel's class are in rows. Look at this picture of where Chantel and her friends sit. Then do Numbers 7 through 10.

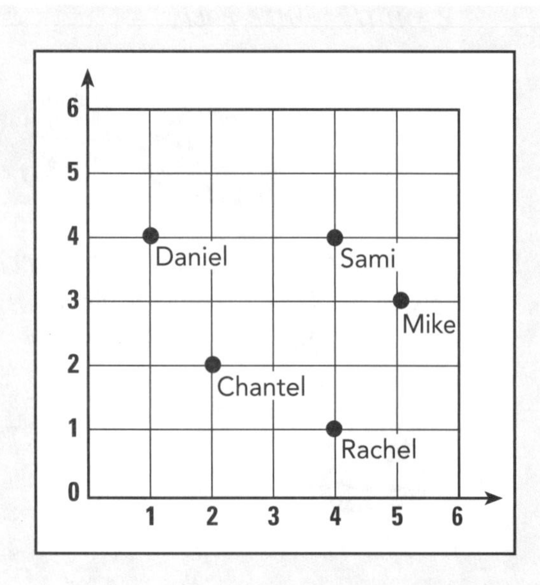

7. Whose desk is over 4 and up 1?

   Ⓐ Daniel's
   Ⓑ Sami's
   Ⓒ Chantel's
   Ⓓ Rachel's

8. Where is Daniel's desk?

   Ⓐ over 4 and up 4
   Ⓑ over 2 and up 2
   Ⓒ over 1 and up 4
   Ⓓ over 4 and up 1

9. Where is Sami's desk?

   Ⓐ over 4 and up 4
   Ⓑ over 5 and up 4
   Ⓒ over 4 and up 3
   Ⓓ over 4 and up 1

10. Whose desk is over 5 and up 3?

    Ⓐ Mike's
    Ⓑ Chantel's
    Ⓒ Daniel's
    Ⓓ Sami's

Read about Chantel's scout troop. Then do Numbers 11 through 14.

Troop 102 Welcomes You

Welcome to Troop 102. We will be doing many fun and important things this year. We will have our meetings on Tuesdays. The second Tuesday in each month will be arts-and-crafts day. We'll visit people and places—and we'll do much, much more. See you soon!

OCTOBER

SUN	MON	TUES	WED	THURS	FRI	SAT
				1	2	3
4	5	6	7	8	9	10
11	12	13	14	15	16	17
18	19	20	21	22	23	24
25	26	27	28	29	30	31

11. What will be the date of arts-and-crafts day in October?

Ⓐ October 14

Ⓑ October 6

Ⓒ October 13

Ⓓ October 20

12. Chantel's troop is holding a bake sale on October 3. What day is this?

Ⓐ Thursday

Ⓑ Saturday

Ⓒ Friday

Ⓓ Sunday

13. On which of these dates will Chantel's troop have a meeting?

Ⓐ October 5

Ⓑ October 14

Ⓒ October 19

Ⓓ October 27

14. The troop will visit a senior center on the third Sunday of the month. What will be the date of the visit?

Ⓐ October 25

Ⓑ October 18

Ⓒ October 11

Ⓓ October 4

PART FIVE: Prepare for a Test

- A test question about graphs may ask for information from a pictograph or a bar graph.
- A test question about charts may ask for information from a Venn diagram.
- A test question about graphs may ask for information from a grid.
- A test question about charts may ask for information from a calendar.

Read about Chantel's bird-watching. Then do Numbers 15 and 16.

Bird-watching

Chantel always keeps her family's bird feeders full of seed. She loves to watch the different kinds of birds that visit the feeders. Chantel sometimes notices a bird that she has never seen before. So she looks in her bird book to see what kind it is. One day she saw 12 cardinals!

Interpreting Graphs and Charts

15. The story tells how many cardinals Chantel saw in one day. Chantel put this information in a pictograph.

If each [bird] stands for 2 birds, which picture shows how many cardinals she saw?

Ⓐ 12 birds

Ⓑ 4 birds

Ⓒ 6 birds

Ⓓ 5 birds

Interpreting Graphs and Charts

16. Chantel made this graph to show all the birds she saw in her backyard yesterday.

How many blue jays did she see?

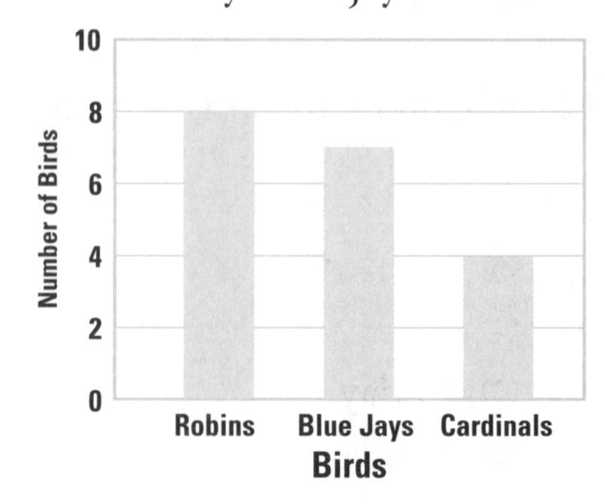

Ⓐ 7 blue jays Ⓒ 8 blue jays

Ⓑ 6 blue jays Ⓓ 4 blue jays

The map shows Chantel's neighborhood. Study the map. Then do Numbers 17 and 18.

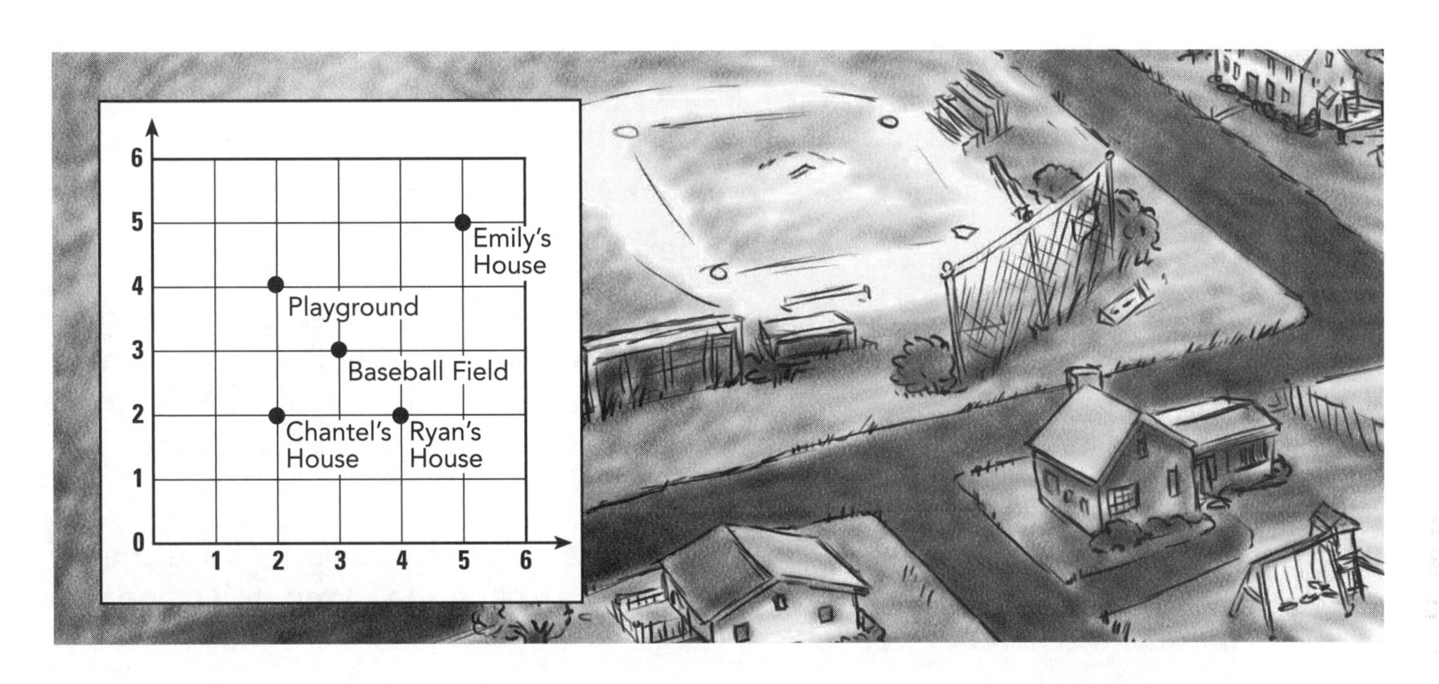

Interpreting Graphs and Charts

17. Which of these locations is over 4 and up 2?

Ⓐ Chantel's house

Ⓑ playground

Ⓒ Ryan's house

Ⓓ baseball field

Interpreting Graphs and Charts

18. During the summer, the people in Chantel's neighborhood have block parties. The parties are on the last Saturday of the month.

What is the date of the block party this month?

AUGUST

SUN	MON	TUES	WED	THURS	FRI	SAT
		1	2	3	4	5
6	7	8	9	10	11	12
13	14	15	16	17	18	19
20	21	22	23	24	25	26
27	28	29	30	31		

Ⓐ August 31

Ⓑ August 26

Ⓒ August 19

Ⓓ August 27

PART ONE: Read a Story

Read the story about Bart and Victoria at the arcade. Then do Numbers 1 through 6.

Wonderland

Bart and Victoria had a fun day at the arcade. They played games and won some prizes. Bart's favorite game is the Wonder Wheel. The game has a circle of lights, and each light has a number. When the wheel comes to a stop, only one light shines. The player wins the number of tickets shown on the light.

On the Wonder Wheel, 10 lights have the number 1; 8 lights have the number 2; 6 lights have the number 5; 4 lights have the number 10; and 1 light has the number 50.

Victoria played the Wonder Wheel four times. She won 5 tickets in the first game and 1 ticket in the second game. Victoria then won 1 ticket in the third game and 1 ticket in the fourth game. Bart and Victoria turned in their tickets for prizes before they went home.

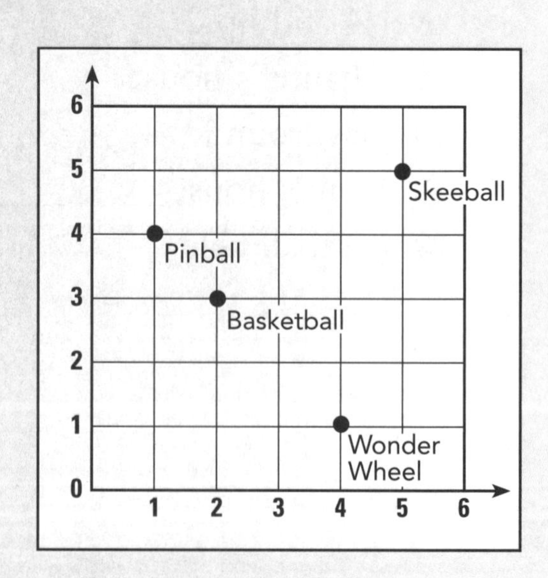

Using Geometry

1. Bart walked around the Mystery Box and counted all of the faces he could see.

How many faces did he count?

Ⓐ 2 faces
Ⓑ 4 faces
Ⓒ 5 faces
Ⓓ 8 faces

Using Geometry

2. Bart also counted all of the triangles he could see on the front of the Mystery Box.

How many triangles did he count?

Ⓐ 6 triangles
Ⓑ 5 triangles
Ⓒ 10 triangles
Ⓓ 4 triangles

Determining Probability and Averages

3. At the Wonder Wheel, Victoria won 5 tickets, then 1 ticket, 1 ticket, and 1 ticket.

What is the average number of tickets that she won in a game?

Ⓐ 5 tickets
Ⓑ 8 tickets
Ⓒ 3 tickets
Ⓓ 2 tickets

Determining Probability and Averages

4. The story tells how many lights there are for each number on the Wonder Wheel.

On what number is the wheel most likely to stop?

Ⓐ 1
Ⓑ 2
Ⓒ 5
Ⓓ 10

Interpreting Graphs and Charts

5. Look at the grid. What game is at the point that is located over 1 and up 4?

Ⓐ Skeeball
Ⓑ Wonder Wheel
Ⓒ Basketball
Ⓓ Pinball

Interpreting Graphs and Charts

6. Bart went to the arcade on October 11.

What day of the week was October 11?

October

Sun.	Mon.	Tues.	Wed.	Thur.	Fri.	Sat.
			1	2	3	4
5	6	7	8	9	10	11
12	13	14	15	16	17	18

Ⓐ Thursday
Ⓑ Friday
Ⓒ Saturday
Ⓓ Sunday

Read the article from a magazine about riddles and puzzles. Then do Numbers 7 through 12.

Puzzle Page

Do you like to solve riddles? Are you crazy for puzzles? If so, try to answer the questions on this page.

Fold a piece of paper. Then punch two holes in the paper. What will the paper look like when it is unfolded?

I am an even number. I am in both circle *A* and circle *B*. What number am I?

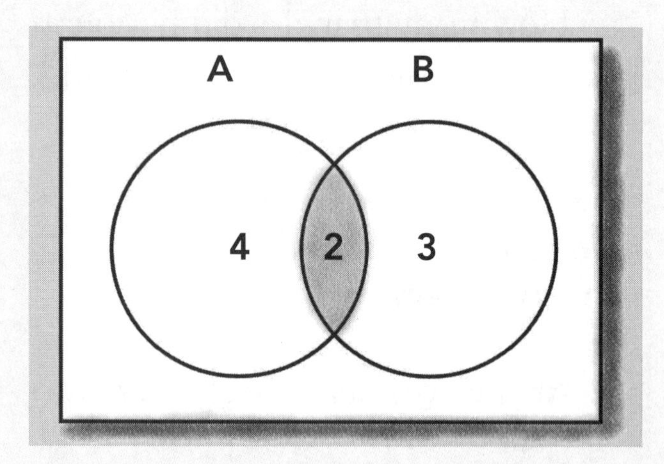

Person	Number of Bounces
Amy	★ ★ ★ ★ ★ ★
Bob	★ ★ ★
Cam	★ ★ ★ ★

★ = 1 bounce

Three friends have a contest. They bounce a ball off their feet and legs. The ball cannot touch the ground. The pictograph shows how many times each friend bounced the ball before it hit the ground. Who won the contest?

Using Geometry

7. Look at the folded piece of paper.

What will the paper look like when it is unfolded?

Ⓐ 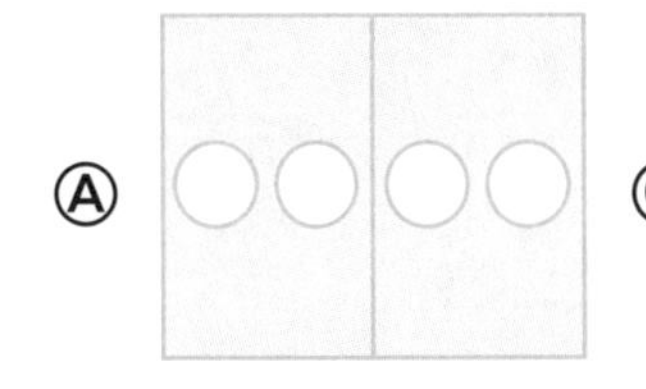

Ⓒ 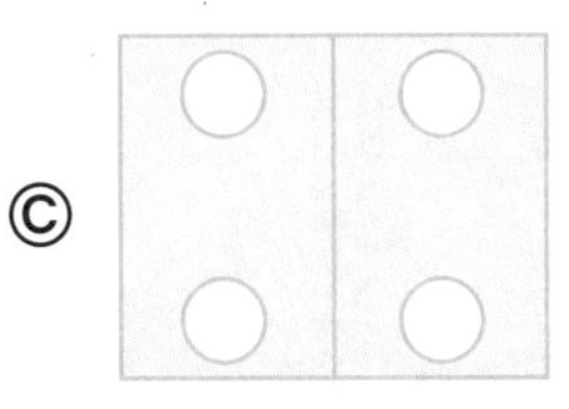

Ⓑ 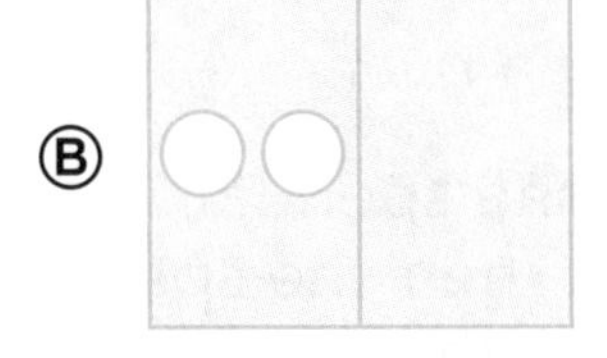

Ⓓ 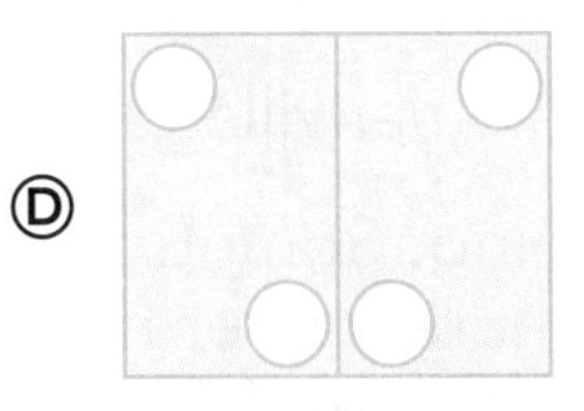

Determining Probability and Averages

10. This is a target for a game played with a ball. A player throws the ball at the target 20 times.

What number would the player probably hit most often?

10	5	20	5
25	5	10	5

Ⓐ 25 Ⓒ 10
Ⓑ 20 Ⓓ 5

Using Geometry

8. Two sides of a piece of paper are each 3 inches long. The other two sides are 5 inches long.

What is the distance around the piece of paper?

Ⓐ 2 inches
Ⓑ 8 inches
Ⓒ 16 inches
Ⓓ 10 inches

Interpreting Graphs and Charts

11. Look at the diagram of circles *A* and *B*. Circle *A* shows the number of boys who have only a bike. Circle *B* shows the number of boys who have only a skateboard.

How many boys have both a bike and a skateboard?

Ⓐ 4 boys Ⓒ 3 boys
Ⓑ 2 boys Ⓓ 9 boys

Determining Probability and Averages

9. Look at the three numbers inside circles *A* and *B*.

What is the average of the numbers?

Ⓐ 9 Ⓒ 2
Ⓑ 6 Ⓓ 3

Interpreting Graphs and Charts

12. Look at the pictograph.

How many more times did Amy bounce the ball than did Cam?

Ⓐ 6 more times
Ⓑ 3 more times
Ⓒ 2 more times
Ⓓ 4 more times

Strategies One–Twelve

FINAL REVIEW

PART ONE: Read a Fable

Read the retelling of an ancient fable. Then do Numbers 1 through 12.

The Big Race

Turtle and Rabbit lived in a small animal village near a river.

Turtle took her time doing everything. Sometimes, she spent all morning crawling out of the river and up the muddy river bank. Then she spent all afternoon crawling down the muddy bank and back into the river. Turtle didn't mind being slow. It was her way, and that was fine with her.

Rabbit, on the other hand, rushed, rushed, rushed. He rushed here, there, and everywhere. He rushed around shouting, "I am faster than anyone!"

One day, Turtle said, "I believe I could beat you in a race."

"Impossible!" Rabbit replied. The other animals agreed. Turtle was way too slow to beat Rabbit. Even so, when Rabbit and Turtle agreed to race, the other animals all hoped Turtle would win.

The morning of the race, Turtle and Rabbit took off from the starting line at the same time. Well, Rabbit took off. As far as anyone could tell, Turtle hardly moved. In fact, after 45 minutes, she had gone only 9 feet. But, both she and Rabbit were on their way.

Rabbit was 2,671 feet ahead of Turtle when he began to yawn. All that rushing here, there, and everywhere had made him quite tired. He knew Turtle would never catch him. So he stopped for a nap.

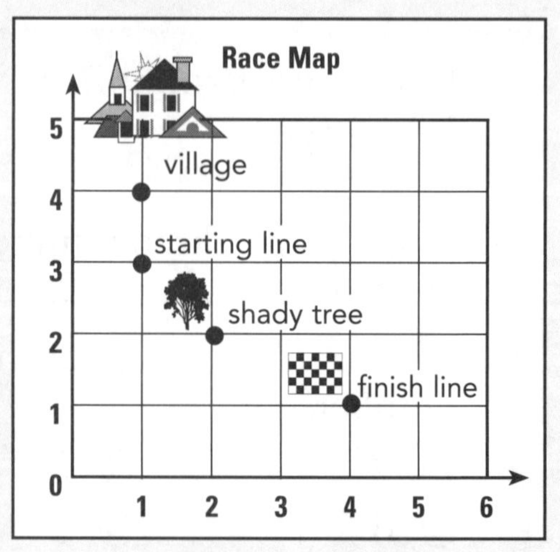

Now, Turtle was slow for sure. But she never stopped—not even when she saw Rabbit sleeping under a shady tree. She never stopped until she crossed the finish line.

Rabbit woke up as the sun was setting. With a stretch and a smile, he raced toward the finish line. But by then, it was too late. Turtle had won the race.

Building Number Sense

1. Rabbit was 2,671 feet ahead of Turtle when he stopped to nap.

What is the place value of the 7 in 2,671?

Ⓐ 7 zeros

Ⓑ 7 ones

Ⓒ 7 tens

Ⓓ 7 hundreds

Applying Subtraction

4. Look at Turtle's practice chart in Number 3.

How many more yards did she crawl on Tuesday than on Monday?

Ⓐ 17 yards

Ⓑ 15 yards

Ⓒ 27 yards

Ⓓ 6 yards

Using Estimation

2. Turtle moved less than 108 inches in the first 45 minutes of the race.

What is the nearest ten of 108?

Ⓐ 8 Ⓒ 90

Ⓑ 110 Ⓓ 100

Applying Multiplication

5. The village animals lined up in 4 rows to watch the race. Each row had 6 animals.

How many animals watched the race?

Ⓐ 30 animals Ⓒ 42 animals

Ⓑ 28 animals Ⓓ 24 animals

Applying Addition

3. Look at how many yards Turtle crawled to practice for the race.

What was the total number of yards she crawled for practice?

Monday	19 yards
Tuesday	36 yards
Wednesday	21 yards

Ⓐ 62 yards Ⓒ 66 yards

Ⓑ 74 yards Ⓓ 76 yards

Applying Division

6. It took Turtle 45 minutes to go 9 feet.

If Turtle moved at a steady pace, how many minutes did it take Turtle to move 1 foot?

Ⓐ 7 minutes

Ⓑ 5 minutes

Ⓒ 9 minutes

Ⓓ 6 minutes

Converting Time and Money

7. The race started at the time shown on the clock. Turtle went 9 feet in 45 minutes.

   What time was it by the time Turtle had gone 9 feet?

   Ⓐ 10:00 A.M.

   Ⓑ 9:30 A.M.

   Ⓒ 8:45 A.M.

   Ⓓ 9:15 A.M.

Using Geometry

10. Look at the shape of each roof on the houses in the village.

    What is the name of this figure?

    Ⓐ circle

    Ⓑ square

    Ⓒ triangle

    Ⓓ cube

Converting Customary and Metric Measures

8. Before the race, Turtle and Rabbit were measured. Standing on his hind legs, Rabbit is 5 inches taller than 1 foot.

   How tall is Rabbit?

   12 inches = 1 foot

   Ⓐ 15 inches Ⓒ 6 inches

   Ⓑ 17 inches Ⓓ 7 inches

Determining Probability and Averages

11. In a box, Turtle has 3 green hats, 7 blue hats, 2 red hats, and 2 yellow hats.

    If she closes her eyes and picks a hat, what color hat will she most likely pick?

    Ⓐ green hat

    Ⓑ blue hat

    Ⓒ red hat

    Ⓓ yellow hat

Using Algebra

9. The race lasted 14 hours. Rabbit ran for a total of 3 hours.

   Which of these number sentences will help you figure out how many hours Rabbit napped?

   Ⓐ $14 - 3 = \square$ Ⓒ $14 + 3 = \square$

   Ⓑ $\square - 14 = 3$ Ⓓ $14 - 14 = \square$

Interpreting Graphs and Charts

12. Look at the map of the race.

    What location is over 1 and up 3?

    Ⓐ shady tree

    Ⓑ finish line

    Ⓒ village

    Ⓓ starting line

PART TWO: Read an Interview

Julio read facts about his favorite children's author. Then he wrote a pretend interview, using the facts he learned. Read the interview. Then do Numbers 13 through 24.

JULIO:	Is Dr. Seuss your real name?
DR. SEUSS:	No, my real name is Ted Geisel. "Seuss" was my mother's last name before she married my father.
JULIO:	Your first book, *And to Think That I Saw It on Mulberry Street*, came out in 1937. Why did you write that book?
DR. SEUSS:	My wife and I were on a ship. I made up a rhyme that went with the sound of the ship's engine. My wife said that I should turn my rhyme into a children's book. So I did.
JULIO:	Did you have any trouble getting your book published?
DR. SEUSS:	Oh, yes! I don't quite remember how many publishers sent the book back to me. I got it back at least 27 times before someone finally agreed to publish it.
JULIO:	How many children's books have you written in all?
DR. SEUSS:	Let's see, including *Green Eggs and Ham, The Cat in the Hat,* and *One Fish, Two Fish,* the total comes to 44.
JULIO:	Is it true that you wrote one of your books on a dare?
DR. SEUSS:	Well, a friend said he would give me $50.00 if I wrote a book using only 50 words. I did. The book was *Green Eggs and Ham.*
JULIO:	My favorite is *The Cat in the Hat*. Can you tell me how you wrote that book?
DR. SEUSS:	Yes. In 1954, there was a report that children's books were dull. My publisher sent me a list of four hundred words that he thought were important. He told me to cut the list to two hundred fifty words and use them all at least once in a book. I used two hundred twenty of them to write *The Cat in the Hat.*
JULIO:	Thank you for your time, Dr. Seuss. And thank you for your books.
DR. SEUSS:	It was my pleasure, Julio.

Dr. Seuss was born in 1904. He died in 1991.

Building Number Sense

13. What number shows how many different words Dr. Seuss used from the list to write *The Cat in the Hat*?

Ⓐ 2,120 Ⓒ 22
Ⓑ 202 Ⓓ 220

Using Estimation

14. Julio estimated how many Dr. Seuss books he has read.

About how many Dr. Seuss books has he read?

Year	Number of Books
2004	12
2003	9
2002	7

Ⓐ 10 books Ⓒ 30 books
Ⓑ 20 books Ⓓ 40 books

Applying Addition

15. Julio counted the number of words on three pages of *Green Eggs and Ham.*

How many words did Julio count?

One page had 8 words.
One page had 6 words.
One page had 7 words.

Ⓐ 21 words Ⓒ 31 words
Ⓑ 20 words Ⓓ 29 words

Applying Subtraction

16. Julio has read 28 Dr. Seuss books.

How many of the 44 Dr. Seuss books has Julio *not* read?

Ⓐ 24 books Ⓒ 16 books
Ⓑ 22 books Ⓓ 12 books

Applying Multiplication

17. Julio decided to write a book called *Dog in a Fog.* There are 6 pages in Julio's book. Each page has 9 words.

How many words are in Julio's book?

Ⓐ 54 words
Ⓑ 45 words
Ⓒ 63 words
Ⓓ 36 words

Applying Division

18. The clerk at the bookstore sold the same number of Dr. Seuss books to 8 customers. In all, he sold 32 books.

How many books did each customer buy?

Ⓐ 6 books
Ⓑ 7 books
Ⓒ 5 books
Ⓓ 4 books

Converting Time and Money

19. The clocks show when Julio started and stopped writing.

How long did Julio write?

Start Stop

Ⓐ 1 hour

Ⓑ 3 hours

Ⓒ 2 hours and 20 minutes

Ⓓ 4 hours and 10 minutes

Converting Customary and Metric Measures

20. Imagine that the cat in *The Cat in the Hat* filled his hat with 4 gallons of water.

How many quarts of water would be in the cat's hat?

4 quarts = 1 gallon

Ⓐ 8 quarts Ⓒ 16 quarts

Ⓑ 12 quarts Ⓓ 1 quart

Using Algebra

21. Julio drew a design made of hats. The missing number in the pattern below is the same as the number of hats he drew.

How many hats did Julio draw?

4, 7, ___, 13, 16

Ⓐ 4 hats Ⓒ 10 hats

Ⓑ 13 hats Ⓓ 11 hats

Using Geometry

22. Julio walked around a stack of 5 Dr. Seuss blocks.

How many faces could he see?

Ⓐ 25 faces

Ⓑ 21 faces

Ⓒ 22 faces

Ⓓ 20 faces

Determining Probability and Averages

23. Julio read for 2 hours Monday, 3 hours Tuesday, 4 hours Thursday, and 3 hours Friday.

What is the average number of hours he read each day?

Ⓐ 6 hours Ⓒ 10 hours

Ⓑ 12 hours Ⓓ 3 hours

Interpreting Graphs and Charts

24. On the first Friday of May, Julio read his pretend interview.

On what date did Julio read?

Ⓐ May 6

Ⓑ May 12

Ⓒ May 5

Ⓓ May 1

May

Sun.	Mon.	Tues.	Wed.	Thur.	Fri.	Sat.
	1	2	3	4	5	6
7	8	9	10	11	12	13
14	15	16	17	18	19	20
21	22	23	24	25	26	27
28	29	30	31			

PART THREE: Read a Story

Read the story about a school project. Then do Numbers 25 through 36.

Ms. Tan shared these facts about the desert, the tundra, and the rain forest with her second-grade class.

Desert: Some deserts are hot and dry. Some are cold and dry. Most plants and animals cannot live well in the desert. That's because there is very little water. Also, the temperature is too hot or too cold. Even so, cactus plants, lizards, snakes, and some other animals do live well in the desert.

Arctic Tundra: The tundra is a very cold, very large place near the Arctic Ocean. Trees do not grow in the tundra. That's because the soil underground is always frozen. But the tundra is home to such animals as the polar bear and the Arctic fox.

Tropical Rain Forest: A tropical rain forest is a warm, rainy place where millions of plants and animals live. In fact, more different kinds of plants and animals live in tropical rain forests than anywhere else on earth.

Ms. Tan asked groups of students to do a report about one of these places. She also asked them to make posters. Liam's group chose to do a report on a tropical rain forest. Here is the poster that the group made.

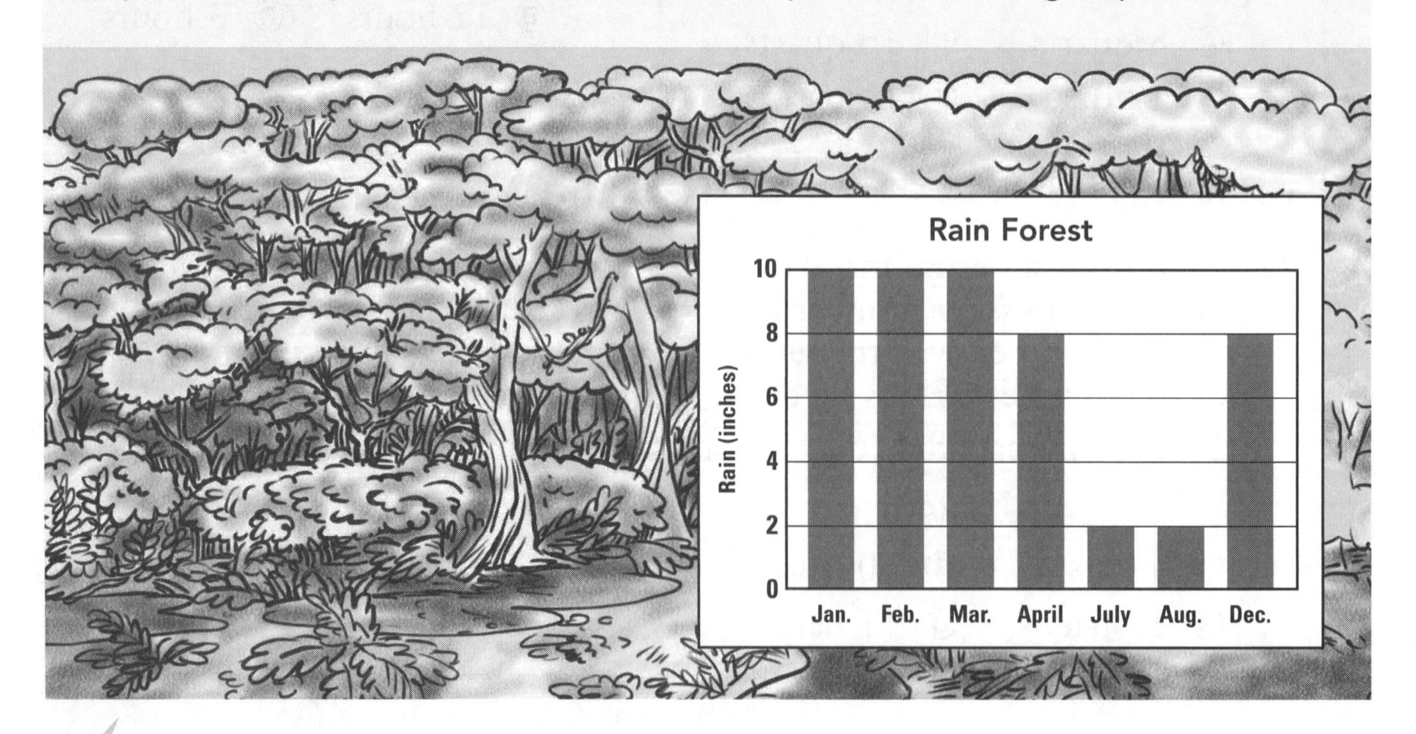

Building Number Sense

25. What problem shows how much rain falls in a rain forest in February and August?

Ⓐ 2 + 2 = 4

Ⓑ 10 + 8 = 18

Ⓒ 10 + 2 = 12

Ⓓ 8 + 8 = 16

Using Estimation

26. A tree could grow to 164 feet in the rain forest.

What is the nearest hundred of 164?

Ⓐ 100

Ⓑ 200

Ⓒ 300

Ⓓ 160

Applying Addition

27. One area of a rain forest has 37 kinds of trees. Another area has 21 kinds of trees.

How many kinds of trees in all are in both areas?

Ⓐ 68 kinds of trees

Ⓑ 56 kinds of trees

Ⓒ 49 kinds of trees

Ⓓ 58 kinds of trees

Applying Subtraction

28. Liam worked for 4 hours on his group's poster. Ana worked for 12 hours on the poster.

How many more hours did Ana work than Liam?

Ⓐ 8 hours Ⓒ 6 hours

Ⓑ 2 hours Ⓓ 9 hours

Applying Multiplication

29. The rain-forest graph shows that it rains 10 inches a month in January, February, and March.

How much rain falls altogether in those 3 months?

Ⓐ 13 inches Ⓒ 33 inches

Ⓑ 60 inches Ⓓ 30 inches

Applying Division

30. Liam saw a picture of a trail of ants marching along the rain-forest floor. The trail was 96 inches long. There were 3 ants for every inch.

How many ants were there?

Ⓐ 33 ants

Ⓑ 32 ants

Ⓒ 28 ants

Ⓓ 23 ants

Converting Time and Money

31. Liam's group paid for supplies with a $10.00 bill and got 83¢ change.

What group of coins totals 83¢?

Ⓐ

Ⓑ

Ⓒ

Ⓓ

Converting Customary and Metric Measures

32. One kind of rain-forest butterfly, when it spreads its wings, is a little wider than a piece of paper.

Which of these measures might be the butterfly's wing span?

Ⓐ 1 foot Ⓒ 1 inch

Ⓑ 1 meter Ⓓ 1 centimeter

Using Algebra

33. Liam wrote this pattern to show rainfall totals for different rain forests in one year.

Find the pattern. What are the two missing numbers?

76, 82, 88, ___, ___

Ⓐ 64 and 70 Ⓒ 100 and 106

Ⓑ 88 and 94 Ⓓ 94 and 100

Using Geometry

34. Liam's group made a border of leaves and glued the border around the poster. The poster is 30 inches wide and 18 inches long.

How many inches around was the border?

Ⓐ 48 inches

Ⓑ 96 inches

Ⓒ 42 inches

Ⓓ 78 inches

Determining Probability and Averages

35. Look at the graph in the poster.

What is the average number of inches of rain for February, March, April, and December?

Ⓐ 8 inches Ⓒ 9 inches

Ⓑ 10 inches Ⓓ 20 inches

Interpreting Graphs and Charts

36. Look at the graph in the poster.

How many inches of rain usually fall in April and in July?

Ⓐ 8 inches and 2 inches

Ⓑ 10 inches and 8 inches

Ⓒ 10 inches and 2 inches

Ⓓ 8 inches and 10 inches

PART FOUR: Read an Article

Read the article about the camel.
Then do Numbers 37 through 48.

The Camel

A camel is a mammal that lives in the desert. The camel has large eyes on the sides of its head. Each eye has three eyelids. Each eye is protected by long, curly eyelashes that keep out the sand and the dust. A camel's thick, bushy eyebrows shield the eyes from the hot desert sun.

A camel walks within hours of birth. The young camel remains close to its mother until five years of age.

A camel has only one hump on its back. This hump contains a built-in food supply. Most people believe that the hump is a place for storing water. The hump is really a large lump of fat. The fat supplies energy for the animal when food is hard to find.

A camel can travel for a long time in the desert with little food or water. It can walk on soft sand where trucks would get stuck. The animal carries people and heavy loads to places that have no roads. It can carry up to 330 pounds for as long as eight hours. A camel stands about 6 to 7 feet tall and weighs from 550 to 1,500 pounds. Its rope-like tail can be 21 inches long.

Millions of people who live in North Africa and Asia depend on the camel to supply many of their needs.

Building Number Sense

37. In what place in line is the shaded camel?

Ⓐ fourth

Ⓑ sixth

Ⓒ third

Ⓓ fifth

Applying Subtraction

40. What is the difference between a camel that weighs 1,321 pounds and one that weighs 786 pounds?

Ⓐ 435 pounds

Ⓑ 535 pounds

Ⓒ 525 pounds

Ⓓ 425 pounds

Using Estimation

38. A camel can weigh up to 1,500 pounds.

What number below can be rounded to 1,500?

Ⓐ 1,582

Ⓑ 1,551

Ⓒ 1,498

Ⓓ 1,437

Applying Multiplication

41. Ahmed's father loaded 80 pounds of supplies on the back of each of 6 camels.

How many pounds of supplies did the 6 camels carry?

Ⓐ 420 pounds

Ⓑ 360 pounds

Ⓒ 560 pounds

Ⓓ 480 pounds

Applying Addition

39. Three camels weigh 728 pounds, 842 pounds, and 1,061 pounds.

What is the total weight of the three camels?

Ⓐ 2,741 pounds

Ⓑ 2,631 pounds

Ⓒ 2,742 pounds

Ⓓ 2,731 pounds

Applying Division

42. Ahmed placed 48 camels into groups of 4.

How many groups of camels did Ahmed put together?

Ⓐ 10 groups

Ⓑ 13 groups

Ⓒ 11 groups

Ⓓ 12 groups

Converting Time and Money

43. Ahmed has the U.S. coins that are shown below.

What is the value of Ahmed's coins?

Ⓐ 85¢

Ⓑ 63¢

Ⓒ 80¢

Ⓓ 75¢

Converting Customary and Metric Measures

44. A camel stands about 2 meters tall.

What measure below is very close to 2 meters?

Ⓐ $6\frac{1}{2}$ inches

Ⓑ $6\frac{1}{2}$ miles

Ⓒ $6\frac{1}{2}$ feet

Ⓓ $6\frac{1}{2}$ centimeters

Using Algebra

45. Ahmed gathered 8 camels together. He fed 3 of the camels.

Which of these number sentences will help you figure out the number of camels in the group that still need to be fed?

Ⓐ $8 - 3 = \square$ Ⓒ $\square - 8 = 3$

Ⓑ $8 \div 3 = \square$ Ⓓ $8 \div 5 = \square$

Using Geometry

46. The distance around a square camel pen is 52 feet.

What is the measure of one side of the square pen?

Ⓐ 14 feet

Ⓑ 13 feet

Ⓒ 15 feet

Ⓓ 12 feet

Determining Probability and Averages

47. Ahmed measured the tails of 4 camels. Put end to end, the 4 tails totaled 80 inches in length.

What is the average length of each tail?

Ⓐ 40 inches

Ⓑ 20 inches

Ⓒ 30 inches

Ⓓ 10 inches

Interpreting Graphs and Charts

48. Study the pictograph about the camel groups.

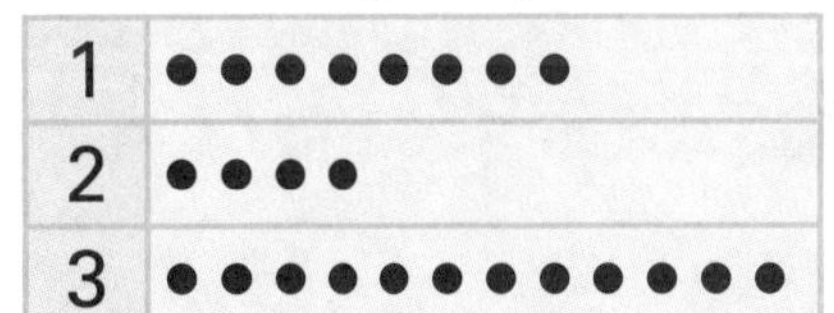

How many camels are in group 3?

Ⓐ 20 Ⓒ 8

Ⓑ 16 Ⓓ 24

Curriculum Associates 4-Step Mathematics Program

Purpose	Series	Objectives
Diagnose	**Comprehensive Assessment of Mathematics Strategies (CAMS® Series)**	• to identify a student's level of mastery for each of 12 math strategies • to develop effective practices in math self-assessment
Teach	**Strategies to Achieve Mathematics Success (STAMS® Series)**	• to provide targeted strategy-specific instruction and practice to students learning key math concepts • to broaden student proficiency in error analysis
Extend	**Extensions in Mathematics™ Series**	• to strengthen students' problem-solving skills and writing skills using graphic organizers • to expand on the 12 standards-based strategies promoted in the ***CAMS® Series*** and ***STAMS® Series***
Assess	**Comprehensive Assessment of Mathematics Strategies II (CAMS® Series II)**	• to assess students' math proficiency at the conclusion of the instructional period • to continue the development to effective practices in math self-assessment